INTERNATIONAL CENTRE FOR MECHANICAL SCIENCES

COURSES AND LECTURES - No. 217

LIMIT ANALYSIS AND RHEOLOGICAL APPROACH IN SOIL MECHANICS

EDITED BY

W. OLSZAK AND L. ŠUKLJE

SPRINGER-VERLAG WIEN GMBH

Originally published by Springer-Verlag Wien New York in 1978

ISBN 978-3-211-81511-3 ISBN 978-3-7091-4352-0 (eBook)

DOI 10.1007/978-3-7091-4352-0

PREFACE

The formal analogy which exists between the basic relationships of the mathematical theory of plasticity and those of soil mechanics, stems for transferring the methods of the first to the wide fields of the second. Of course, the essential physical differences have to be appropriately taken into account. This primarily refers to the notion of the yield criterion as well as to that of the plastic potential which may, in certain cases, be adopted. The above approach is generally being used when investigating problems of limit equilibrium conditions.

However, experimental evidence clearly shows that the elastic-plastic approach can constitute the first approximation to reality only and that soils often exhibit rheological properties. Thus, in order to be able to adequately represent their response to various types of loading programmes, we have to consider both the irreversible and time-dependent character.

These facts are well reflected in the papers of the present volume. It contains four series of lectures on some soil mechanics problems, presented at the International Centre for Mechanical Sciences within the scope of the Summer Session in 1976. Three series are related to the theory of plasticity as applied to limit equilibrium problems, while the fourth deals with the influence of non-linear viscous response onto the development of stresses and strains in solids.

The classical limit equilibrium analysis is based on the system of differential equations obtained by combining the appropriate yield criterion with the equilibrium conditions. Using Mohr-Coulomb's failure law, the joint differential equations were first deduced, for plane wedge-shaped plastic domains, by Boussinesq (1885), and in a more general form by Kötter (1903). The use of these equations was very limited until appropriate numerical procedures facilitated their wide application to bearing capacity and earth pressure problems (Caquot 1943, Sokolovskij 1954, and others).

The adaptation of the theory of plasticity to strain-hardening and strain-softening media, with the stress-strain and strain-displacement relationships incorporated in the analysis, resulted in the use of the plasticity theory in the wide domain from the "at rest" up to the failure stress-states. On the other hand, the consideration of the flow rule and of the kinematic relations has permitted to construct the velocity fields corresponding to the limit stress fields of soils considered as rigid-plastic media; so, for statically indeterminate problems the static solutions can be checked by the admissibility of associated kinematic fields. The flow rule has been formulated either from pure kinematic considerations of equal or unequal slidings of free rotating curved rhomboidal elements along the stress characteristics (as presented, in this book, by de Josselin de Jong), or by the derivation from a plastic potential (as accepted by Mróz and Salençon). The first concept leads to the conclusion that the principal directions of stress and strain-velocity tensors need not coincide (de Josselin de Jong), and the experimental evidence for friction soils proves that the plastic potential surfaces only exceptionally coincide with the yield surfaces.

The consequences of the above stated plastic soil behavior for the interpretation of the limit stress anlysis are the main subject of the lectures. The discussions are related to the Mohr-Coulomb failure law, alternatively also to the yield criteria of Huber-von Mises and of Tresca. – Particularly, de Josselin de Jong presents the problems of discontinuities in stress fields and of the non-coaxiality of stress and strain-velocity tensors. – With the assumption that the plastic potential is a singular vertex lying on the Mohr-Coulomb failure surface and obtained by the intersection of three planes, Mróz discusses the solutions corresponding to different flow rules for plane deformations : the associated as well as the non-associated, the coaxial and non-coaxial flow rules whereby the rate of dilatation is either proportional to the shear strain rate or variable, and, finally, flow rules affected by the stress rate. –Salençon presents a review of the application of the theory of plasticity in soil mechanics and, after having introduced the rigid-plastic scheme by a presentation of the successive yielding of elasto-plastic media, discusses the limit analysis by taking into account the non-homogeneity of soil media obeying either the associated or non-associated flow rules, in either plane-strain or axially symmetric conditions. Concluding his systematic considerations, he presents a critical review of the combined static and kinematic limit analysis as to the reliability and applicability of the results obtained.

For forecasting the development of stresses and strains in saturated soils exhibiting non-viscous linear deformability, Biot's equation (1935, 1941) can efficiently be used. It combines the equilibrium equations for elastic media and the diffusion equation for isotropic, homogeneous, saturated soils. Appropriate numerical procedures facilitate the solution of the equation for arbitrary boundary conditions. Linear visco-elastic models when appropriately transformed into elastic models, can be treated by similar procedures.

For non-linear viscous, saturated or partly saturated soils, the joint solution is very complicated. In the fourth series of lectures presented in this book, Šuklje has brought to light the effects of non-linear viscous properties by separating the analysis of stresses and strains, when excluding the influence of the seepage resistance, from the solution of the diffusion equation for total stress states which are or which are supposed to be statically determinate. The rheological relationships used in the analysis are of hypoelastic character and are expressed in terms of octahedral values of stresses and strains; they are stress-path dependent and valid for monotonically increasing stresses and stress levels only. The combination of the two separate analyses for statically indeterminate total stress conditions involves some simplifying assumptions.

By the joint solution in which the real soil properties would be considered in a form as perfect as possible and in which the condition of small strains would not be imperative, the forecasting of the displacements in soils could be considerably improved. As to the bearing capacity of soils, the criterion of admissible displacements and of progressive yielding could then successfully replace the limit equilibrium criterion. Now, the great number of physical parameters appearing in general concepts and requiring cumbersome experimental work, furthermore the multi-layer, often lens-shaped composition of soils, the difficulties in programming numerical computations, the capacity of computers and, last not least, the costs of performing the computations restrict the realization of general concepts. Thus, the adaptation of the soil model to the particular problem to be solved, as well as separations and simplifications in formulating and in elaborating the solutions remain justified also in future investigations of geotechnical problems.

W. Olszak *L. Šuklje*

December 1977.

CONTENTS

LIST OF CONTRIBUTORS

J. DE JONG – Department of Civil Engineering, Delft University of Technology, DELFT, The Netherlands.

Z. MRÓZ and Cz. SZYMÁNSKI – Institute of Fundamental Technological Research, PAN, WARSZAWA, Poland.

J. SALENCON – Laboratoire de Mécanique des Solides, Eole Politechnique, Ecole Nationale Superieure des Mines de Paris, PALAISEAU, France.

L. ŠUKLJE – Fakulteta za Arhitekturo, Gradbenivšto in Geodezijo, Univerza v Ljubljani, LJUBLJANA, Yugoslavia.

MODEL FOR THE BEHAVIOUR OF GRANULAR MATERIALS IN PROGRESSIVE FLOW

G. de Josselin de Jong

Introduction

These lecture notes are concerned with the behaviour of granular assemblies, possessing internal friction and cohesion, in the limit state of stress at progressive flow.

The first part deals with the stress distribution in the limit state. The equations along stress characteristics are developed in sections 1.1 and 1.2. Boundary conditions admitting the construction of a stress field are discussed in section 1.3.

In section 1.4 special attention is paid to the possibility, that discontinuities originate in the interior of a continuous field of limiting strèsses. This situation is encountered when two stress characteristics of the same family intersect. It is shown, that the field can be extended beyond the intersection point by introducing an intersection of the conjugate stress characteristics. The strength and direction of the discontinuity emanating from the common intersection point is established.

The second part deals with the model for the strainrate behaviour of granular assemblies in progressive flow, proposed by the author (1958, 1959, 1971, 1977 a, b). This model was called the double sliding, free rotating model, because it consists of slices, that can slide with respect to each other in the two conjugate stress characteristic directions, and that are free to rotate.

The mechanical behaviour of the model is expressed in terms of the sliding rates a,b and the structural rotation Ω in section 2.1. In section 2.2. relations are given for a,b and Ω in terms of the velocity components in the curvilinear s^1, s^2 –coordinate system corresponding to the stress characteristics. In section 2.3 a few properties of the curvilinear coordinates are considered, necessary for developing in section 2.4 restrictions imposed on the derivatives of a,b and Ω in order to satisfy compatibility. The bending character of these relations is discussed in section 2.5.

In section 2.6 the relations for a,b and Ω are expressed in cartesian velocity components, in order to establish the constitutive relations valid at a point. In sections 2.7 and 2.8 special properties of the model concerning rotation, noncoaxiality and lack of normality are discussed.

The use of the constitutive relations in solving boundary value problems is elaborated in section 2.9. Relations are established between the direction of a line and its change of length. With these relations it is possible to verify whether a boundary condition, expressed in terms of velocities, is acceptable. It is shown how to construct the velocity in a point of the field compatible with velocities in adjacent points. Instead of one unique velocity, the solution turns out to consist of a range of possibilities.

The objective of this presentation is not to review all existing theories concerning the behaviour of granular media in the limit flow state, but to describe the model proposed by the author. In order to obtain a consistent development of the relevant relations the theory of stress characteristics was added in Chapter 1. This theory has been treated by many authors, see f.i. Sokolovskii (1965). The treatment of a stress discontinuity starting internally as described in section 1.4. and the greater part of the chapter 2 on kinematics were not published previously.

1. Limit Stress State

1.1. Differential equations describing a limit stress state

In these notes a region of a granular assembly is considered, that is entirely in the state of limit stress, corresponding to a possible state of continuous deformation in progressive flow. The material is assumed to obey the Coulomb-Mohr limiting stress condition. This condition limits the relation between major and minor principal stresses and is assumed to be independent of the magnitude of the intermediate principal stress, for simplicity here.

The plane case is considered with the intermediate principal stress in z direction. The stresses in the x ; y plane are shown in figure 1 left side, where the arrow directions correspond to positive stress values. Every line in the x;y plane stands for a plane perpendicular to x;y through that line.

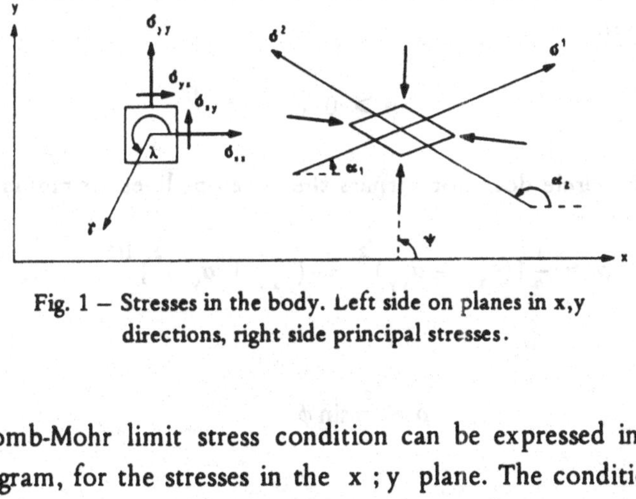

Fig. 1 – Stresses in the body. Left side on planes in x,y directions, right side principal stresses.

The Coulomb-Mohr limit stress condition can be expressed in terms of the Mohr stress diagram, for the stresses in the x ; y plane. The condition is, that the stress circle cannot surpass two envelope lines, shown in figure 2. It is assumed here, that these envelopes are straight lines corresponding to a constant value of the angle of internal friction, ϕ . They intersect in a point O at a distance c cotan ϕ from the origin, when there is cohesion of magnitude c.

Let p be the distance between O and M, the centre of the stress circle, then

$$p = -\frac{1}{2}(\sigma_{xx} + \sigma_{yy}) + c \cot an\phi \,. \qquad (1.1.1.)$$

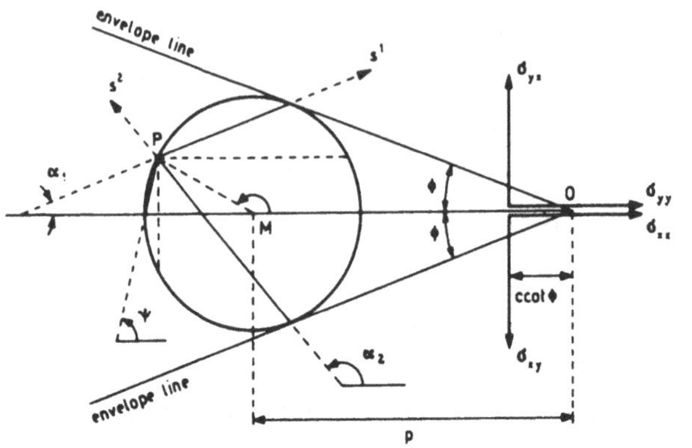

Fig. 2 – Mohr circle for stresses in limiting stress conditions .

In order that the centre lies between the envelope lines, M must be to the left of O, and this limits p to

(1.1.2.) $$p \geqslant 0 \ .$$

In order that the circle does not surpass the envelope lines, its radius ρ defined by

(1.1.3.) $$\rho = \frac{1}{2}[(\sigma_{xx} - \sigma_{yy})^2 + (\sigma_{xy} + \sigma_{yx})^2]^{1/2}$$

must satisfy

(1.1.4.) $$\rho \leqslant p \sin\phi \ .$$

The region considered in these notes is entirely in the limit stress state. For every point of this region the stress circle then touches the envelope lines, such that $\rho = p\sin\phi$, and the circle is as shown in figure 2. On this stress circle the point P is the centre of directions of planes. Let ψ be the angle between x-axis and the plane of major principal compression stress, as indicated in figures 1 and 2. The stresses can then be expressed in terms of p and ψ as

$$\left.\begin{aligned} \sigma_{xx} &= -p + p\sin\phi\cos 2\psi + c\,\mathrm{cotan}\,\phi \\ \sigma_{xy} &= \sigma_{yx} = p\sin\phi\sin 2\psi \\ \sigma_{yy} &= -p - p\sin\phi\cos 2\psi + c\,\mathrm{cotan}\,\phi\,. \end{aligned}\right\} \quad (1.1.5.)$$

When the flow velocities are small enough to disregard inertial forces, static equilibrium requires

$$\left.\begin{aligned} (\partial\sigma_{xx}/\partial x) + (\partial\sigma_{yx}/\partial y) + \gamma\cos\lambda &= 0 \\ (\partial\sigma_{xy}/\partial x) + (\partial\sigma_{yy}/\partial y) + \gamma\sin\lambda &= 0 \end{aligned}\right\} \quad (1.1.6.)$$

where γ is body force per unit volume acting in a direction λ with respect to the x-axis, see figure 1. In order to satisfy equilibrium of moments, the additional requirement

$$\sigma_{xy} = \sigma_{yx} \quad (1.1.7.)$$

exists, when there are no couple stresses.

The expressions (1.1.5.) satisfy (1.1.7.) and introduced in (1.1.6.) give two equations for the two unknowns p; ψ of the form

$$\begin{aligned} &-\frac{\partial p}{\partial x} + \frac{\partial p}{\partial x}\sin\phi\cos 2\psi - 2p\sin\phi\sin 2\psi\frac{\partial\psi}{\partial x} \\ &\quad + \frac{\partial p}{\partial y}\sin\phi\sin 2\psi + 2p\sin\phi\cos 2\psi\frac{\partial\psi}{\partial y} + \gamma\cos\lambda = 0 \end{aligned} \quad (1.1.8.)$$

$$\begin{aligned} &+\frac{\partial p}{\partial x}\sin\phi\sin 2\psi + 2p\sin\phi\cos 2\psi\frac{\partial\psi}{\partial x} \\ &-\frac{\partial p}{\partial y} - \frac{\partial p}{\partial y}\sin\phi\cos 2\psi + 2p\sin\phi\sin 2\psi\frac{\partial\psi}{\partial y} + \gamma\sin\lambda = 0\,. \end{aligned} \quad (1.1.9.)$$

1.2. Quasi-linear hyperbolic character of the differential equations

The differential equations (1.1.8.), (1.1.9.) are quasi-linear, because they consist of first order partial derivatives combined with coefficients, which contain the unknowns p; ψ . The system can be shown to be hyperbolic by bringing the equations in characteristic normal form and determining the eigen values of the coefficient matrix. There are two different, real eigenvalues and the corresponding characteristic directions have angles α_1, α_2 with the x-axis given by

(1.2.1.)
$$\begin{cases} \alpha_1 = \psi - \frac{1}{4}\pi - \frac{1}{2}\phi \\ \alpha_2 = \psi + \frac{1}{4}\pi + \frac{1}{2}\phi . \end{cases}$$

Planes through the lines in these directions are acted upon by stresses corresponding to the tangent points of stress circle and envelope lines.

The system of equations can be transformed in the characteristic direction by making the following combinations. Multiplying (1.1.9.) by $\cos\alpha_2$ multiplying (1.1.8.) by $\sin\alpha_1$ and subtracting gives

(1.2.2)
$$\begin{aligned} &+ \cos\phi \cos\alpha_1 \frac{\partial p}{\partial x} + 2p \sin\phi \cos\alpha_1 \frac{\partial \psi}{\partial x} + \\ &+ \cos\phi \sin\alpha_1 \frac{\partial p}{\partial y} + 2p \sin\phi \sin\alpha_1 \frac{\partial \psi}{\partial y} + \gamma \sin(\lambda - \alpha_2) = 0 . \end{aligned}$$

Multiplying (1.1.9.) by $\cos\alpha_1$, multiplying (1.1.8.) by $\sin\alpha_1$ and subtracting gives

(1.2.3.)
$$\begin{aligned} &- \cos\phi \cos\alpha_2 \frac{\partial p}{\partial x} + 2p \sin\phi \cos\alpha_2 \frac{\partial \psi}{\partial x} + \\ &- \cos\phi \sin\alpha_2 \frac{\partial p}{\partial y} + 2p \sin\phi \sin\alpha_2 \frac{\partial \psi}{\partial y} + \gamma \sin(\lambda - \alpha_1) = 0. \end{aligned}$$

In this form the equations contain the directional differentiations in the characteristic α_1 ; α_2-directions. Coordinates s^1 ; s^2 can be introduced in these directions respectively, see figure 3. These coordinates form a curvilinear network, that is equiangular because all lines intersect at the same angle of magnitude

(1.2.4.)
$$\alpha_2 - \alpha_1 = \frac{1}{2}\pi + \phi .$$

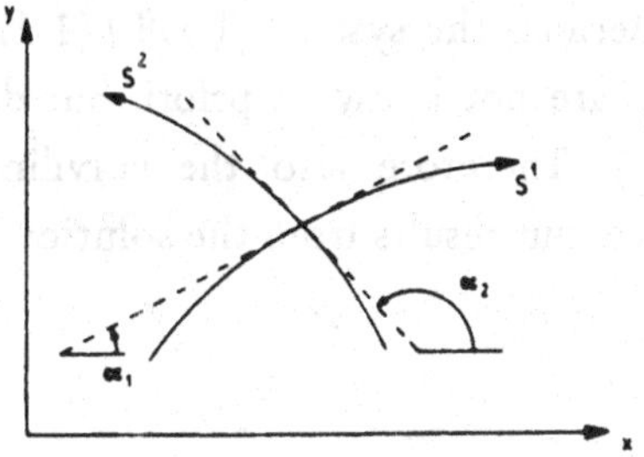

Fig. 3 – Curvilinear stress characteristics, s^1 ; s^2.

In order to use $s^1 ; s^2$ as coordinates, it is necessary to introduce scale parameters $\lambda_1 ; \lambda_2$ representing the absolute value of the base vectors. The relation between the curvilinear, equiangular $s^1 ; s^2$ coordinate system and the cartesian $x ; y$ coordinates is given by the following Jacobi matrix

$$\left.\begin{aligned} \frac{\partial x}{\partial s^1} &= \lambda_1 \cos\alpha_1 & \qquad \frac{\partial y}{\partial s^1} &= \lambda_1 \sin\alpha_1 \\ \frac{\partial x}{\partial s^2} &= \lambda_2 \cos\alpha_2 & \qquad \frac{\partial y}{\partial s^2} &= \lambda_2 \sin\alpha_2 \, . \end{aligned}\right\} \qquad (1.2.5.)$$

Using the chainrule, gives then for any variable f the relation

$$\frac{1}{\lambda_i}\frac{\partial f}{\partial s^i} = \frac{1}{\lambda_i}\left(\frac{\partial f}{\partial x}\frac{\partial x}{\partial s^i} + \frac{\partial f}{\partial y}\frac{\partial y}{\partial s^i}\right) = \frac{\partial f}{\partial x}\cos\alpha_i + \frac{\partial f}{\partial y}\sin\alpha_i \, . \qquad (1.2.6.)$$

Dividing (1.2.2.) and (1.2.3.) through by $2p\sin\phi$ and using this result gives

$$+\frac{1}{2\tan\phi}\frac{\partial(\ln p)}{\lambda_1 \partial s^1} + \frac{\partial\psi}{\lambda_1\partial s^1} + \frac{\gamma}{2p}\frac{\sin(\lambda-\alpha_2)}{\sin\phi} = 0 \qquad (1.2.7.)$$

$$-\frac{1}{2\tan\phi}\frac{\partial(\ln p)}{\lambda_2 \partial s^2} + \frac{\partial\psi}{\lambda_2\partial s^2} + \frac{\gamma}{2p}\frac{\sin(\lambda-\alpha_1)}{\sin\phi} = 0 \, . \qquad (1.2.8.)$$

The equations (1.2.7.) and (1.2.8.) have a hyperbolic form with s^1, s^2 as characteristics, which are usually called the *stress characteristics*. Since differentiation is with respect to s^1 only in (1.2.7.) and s^2 only in (1.2.8.), the equations can be integrated in s^1, s^2-directions respectively.

Because the system (1.1.8.) (1.1.9.) is quasi linear the characteristic directions α_1, α_2 are not known a priori, but depend on the solution of ψ, as indicated by (1.2.1.). Therefore also the curvilinear coordinates mesh s^1 ; s^2 is unknown in advance, but results from the solution of ψ.

1.3. Boundary values necessary for solution

If p and ψ are known on a boundary, Σ_σ, integration of (1.2.6.) and (1,2,7,) can proceed along s^1; s^2-lines that intersect Σ_σ. In the field adjacent to Σ_σ a unique solution exists for p and ψ in every point, where two conjugate stress characteristics meet, that both pass through Σ_σ .

On a boundary not p and ψ are given, but stresses $\sigma_{\eta\eta}$ and $\sigma_{\eta\xi}$ (see figure 4), where $\xi;\eta$ are coordinates along and perpendicular to Σ_σ. The values of p and ψ on the boundary can be found from these stresses by use of equations (1.1.5.) applied to the $\xi;\eta$-system.

$$\sigma_{\eta\eta} = -p - p\sin\phi\cos 2\psi_\Sigma + c\,\mathrm{cotan}\,\phi \qquad (1.3.1.)$$

$$\sigma_{\eta\xi} = +\, p\sin\phi\sin 2\psi_\Sigma\,. \qquad (1.3.2.)$$

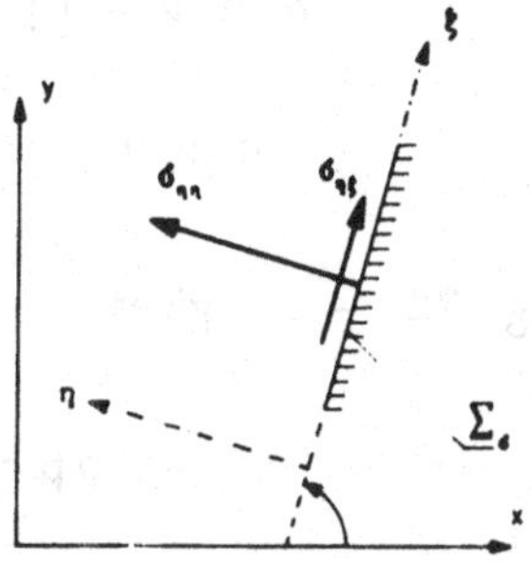

Fig. 4 – Boundary with boundary stresses

If μ is the angle between the boundary Σ_σ and x-axis, then ψ_Σ the angle between the plane of major principal compression stress and ξ-axis is related to ψ by

$$\psi_\Sigma = \psi - \mu\,. \qquad (1.3.3.)$$

In order that the boundary stresses are tolerated by the material, the stresspoint S(see figure 5) corresponding to $\sigma_{\eta\eta}$; $\sigma_{\eta\xi}$ must lie between the envelope lines. This is ensured if in the first place

$$\sigma_{\eta\eta} \leqslant c\,\mathrm{cotan}\,\phi \qquad (1.3.4.)$$

and in the second place

$$|\nu| \leqslant \phi\,, \qquad (1.3.5.)$$

where ν is defined (see figure 5) by

$$\tan \nu = - \sigma_{\eta\xi} / (c \cot\text{an}\, \phi - \sigma_{\eta\eta}) \, . \tag{1.3.6.}$$

In the situation represented in figure 5, both $\sigma_{\eta\eta}$ and $\sigma_{\eta\xi}$ have negative values.

Introduce σ as the length of OS in figure 5, then

$$\begin{aligned} c \cot\text{an}\, \phi - \sigma_{\eta\eta} &= \sigma \cos \nu \\ - \sigma_{\eta\xi} &= \sigma \sin \nu \end{aligned} \tag{1.3.7.}$$

and (1.3.1.) (1.3.2.) are with (1.3.3.)

$$\sigma \cos \nu = p[1 + \sin \phi \cos(2\psi - 2\mu)] \tag{1.3.8.}$$

$$\sigma \sin \nu = - p \sin \phi \sin (2\psi - 2\mu) \, . \tag{1.3.9.}$$

Eliminating $(2\psi - 2\mu)$ gives

$$\sigma^2 - 2\sigma p \cos \nu + p^2 \cos^2 \phi = 0 \tag{1.3.10.}$$

with two solutions

$$\begin{matrix} p_1 \\ p_2 \end{matrix} = (\sigma/\cos^2 \phi) \, [\cos \nu \pm (\sin^2 \phi - \sin^2 \nu)^{1/2}] \, . \tag{1.3.11.}$$

The upper sign corresponds to the larger circle (1) through S and the lower sign to the smaller circle (2), see figure 5.

By using (1.3.11.) in (1.3.8.) and (1.3.9.) there results

$$\sin \phi \cos (2 \begin{matrix} \psi_1 \\ \psi_2 \end{matrix} - 2\mu) = - \sin^2 \nu \mp \cos \nu (\sin^2 \phi - \sin^2 \nu)^{1/2} \tag{1.3.12.}$$

$$\sin \phi \sin (2 \begin{matrix} \psi_1 \\ \psi_2 \end{matrix} - 2\mu) = - \sin \nu \cos \pm \sin \nu (\sin^2 \phi - \sin^2 \nu)^{1/2} \tag{1.3.13.}$$

The relations (1.3.11.) (1.3.12.) (1.3.13.) express p and ψ in terms of σ and ν , which are obtained from $\sigma_{\eta\eta}$, $\sigma_{\eta\xi}$ the boundary stresses by use of (1.3.7.).

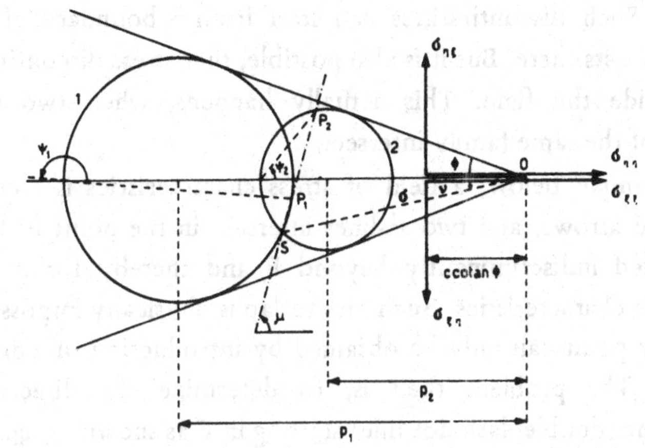

Fig. 5 – Stresspoint S corresponding to boundary stress and the two corresponding limiting stress circles

1.4. Stress discontinuity

Because the hyperbolic system of differential equations is quasilinear, it is possible that discontinuities occur along lines, that deviate from the stress characteristics. Such discontinuities can start from a boundary, if a jump in the variables p; ψ exists there. But it is also possible, that stress discontinuities originate somewhere inside the field. This actually happens, when two different stress characteristics of the same family intersect.

In the example below, a field of stress characteristics is constructed in the direction of the arrows, and two s^1-lines intersect in the point F. In figure 6a the field is extended indiscriminantly beyond F and thereby forms an overlapping pattern of stress characteristics. Such an overlap is physically impossible and unique stresses in every point can only be obtained by introduction of a discontinuity line starting in F. The problem then is, to determine the direction μ of the discontinuity line (double dash-dot line) starting in F as shown in figure 6b.

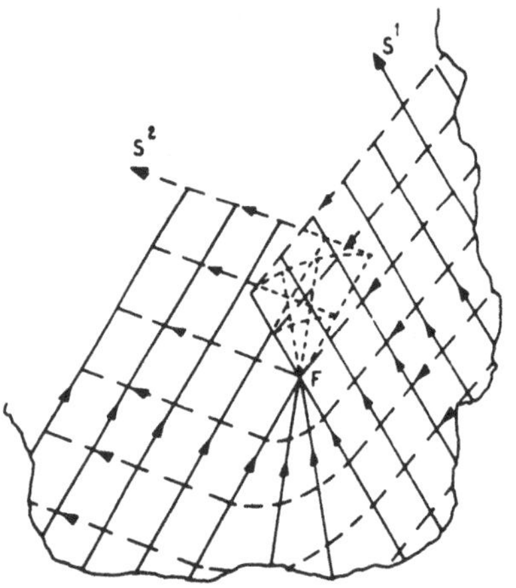

Fig. 6a – Overlapping pattern of indiscriminantly extended field

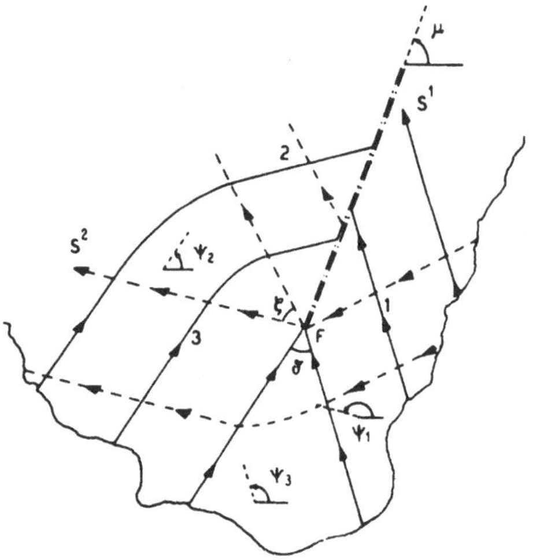

Fig. 6b – Stress discontinuity introduced in F.

The magnitudes of the angles ψ_1 and ψ_3, occurring in the regions (1) and (3) on either side of the s^1 lines, that meet in F are known from the previous constructions. If ϑ is the angle enclosed by the two s^1-lines, we have

$$\vartheta = \psi_1 - \psi_3 . \tag{1.4.1.}$$

Because the bodyforce term in (1.2.8.) can be disregarded, when this relation is integrated along an s^2-line close to F, there results

$$(\ln p_1 - \ln p_3) - 2\tan\phi\,(\psi_1 - \psi_3) = 0 \qquad (1.4.2.)$$

or

$$(p_1/p_3) = \exp[\,2\vartheta \cdot \tan\phi\,]\,. \qquad (1.4.3.)$$

The construction of the field must be continued by introducing in point F a fan of s^2-lines that intersect in F (figure 6b). The fan has an as yet unknown angle of aperture ζ. In the region (2) above this fan, the magnitude of ψ_2 is given by

$$\psi_3 - \psi_2 = \zeta\,. \qquad (1.4.4.)$$

Integration of (1.2.7.) over a s^1-line, through this fan close to F gives

$$(\ln p_3 - \ln p_2) + 2\tan\phi\,(\psi_3 - \psi_2) = 0 \qquad (1.4.5.)$$

or

$$(p_3/p_2) = \exp[\,-\,2\zeta \cdot \tan\phi\,]\,. \qquad (1.4.6.)$$

Combining (1.4.3.) and (1.4.6.) gives

$$(p_1/p_2) = \exp[\,2(\vartheta - \zeta)\tan\phi\,]\,. \qquad (1.4.7.)$$

The value of ζ must be chosen in such a manner, that the stresses in the region (2) are in equilibrium with the stresses in the region (1), across the discontinuity line at the as yet unknown angle μ .

There are two unknowns $(\zeta\,;\mu)$ and two conditions that the two stresses $\sigma_{\eta\eta}\,;\,\sigma_{\eta\xi}$ on either side of the discontinuity line are equal. This is a situation as in figure 5, where one stresspoint S satisfies two circles. Therefore relations

(1.3.11.) (1.3.12.) (1.3.13.) can be used here. They permit to determine ζ as follows.

From (1.3.11.) it is found that

$$\frac{p_1}{p_2} = \frac{\cos\nu + (\sin^2\phi - \sin^2\nu)^{1/2}}{\cos\nu - (\sin^2\phi - \sin^2\nu)^{1/2}} = \frac{[\cos\nu + (\sin^2\phi - \sin^2\nu)^{1/2}]^2}{\cos^2\phi} =$$

$$= \frac{\cos^2\phi}{[\cos\nu - (\sin^2\phi - \sin^2\nu)^{1/2}]^2}$$

and this gives

(1.4.8.) $$\left(\frac{p_1}{p_2}\right)^{1/2} - \left(\frac{p_2}{p_1}\right)^{1/2} = \frac{2(\sin^2\phi - \sin^2\nu)^{1/2}}{\cos\phi} .$$

From (1.3.12) (1.3.13.) it is found that

$$\sin^2\phi\cos 2(\psi_1 - \psi_2) = 2\sin^2\nu - \sin^2\phi$$

and this gives

$$\sin\phi\sin(\psi_1 - \psi_2) = (\sin^2\phi - \sin^2\nu)^{1/2}$$

and with (1.4.1.) (1.4.4.)

(1.4.9.) $$\sin\phi\sin(\vartheta + \zeta) = (\sin^2\phi - \sin^2\nu)^{1/2}.$$

Combining (1.4.7.) (1.4.8) (1.4.9.) gives

(1.4.10.) $$\text{Sinh}\,[\tan\phi\,(\vartheta - \zeta)] = \tan\phi \cdot \sin(\vartheta + \zeta) .$$

This expression (1.4.10.) permits to determine ζ, if ϑ is known. With ζ the stress situation in region (2) can be found in terms of p_2 ; ψ_2 by use of (1.4.4.) and (1.4.7.). Finally μ can be found by use of the relation

(1.4.11.) $$\psi_1 + \psi_2 - 2\mu = \frac{1}{2}\pi - \nu ,$$

which can be deduced from (1.3.12.) and (1.3.13.)

The relation (1.4.10.) was developed by Vermeer (1973) and used in a computer program for determining stress fields with internally originating discontinuities. Figs. 7a, 7b show fields of stress characteristics obtained with this program.

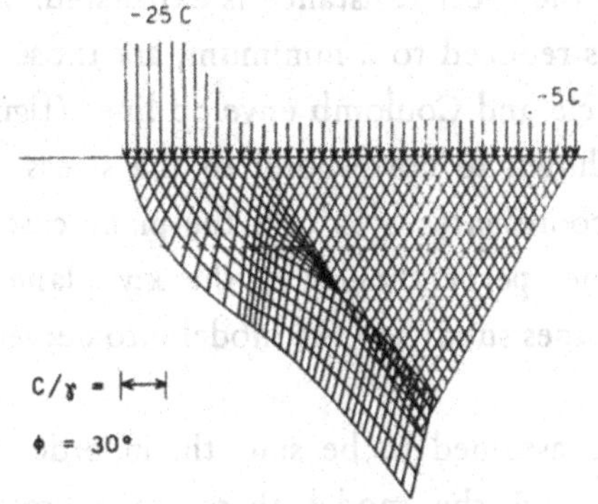

Fig. 7a – Overlapping pattern of indiscriminantly extended field.

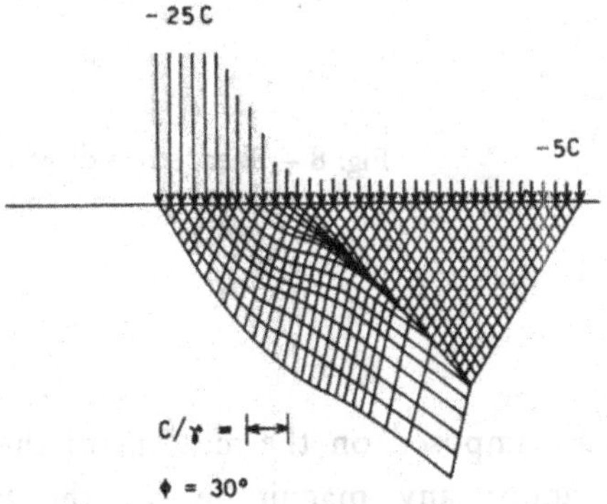

Fig. 7b – Stress discontinuity introduced at intersection point.

2. Double sliding, free rotating model for progressive flow

2.1. Sliding slices model. Definition and properties of a,b, and Ω .

The model proposed by the author (1957, 1959, 1971) for granular media in progressive flow, consists of smooth separation planes coinciding with the stress characteristics, along which sliding occurs exclusively. This model is based on the assumption, that in a real granular medium at limit stress state, sliding will occur along those planes, where the shear resistance is exhausted. The planes, where the avialable shear resistance is reduced to a minimum, are those corresponding to the tangent points of stress circle and Coulomb envelop lines (figure 2). In section 1.2. the tangent points were shown to correspond to the stress characteristics, which were identified with the coordinates $s^1;s^2$. In the plane case considered here, the $s^1;s^2$ lines stand for planes perpendicular to the x;y plane, intersecting the x;y plane along $s^1;s^2$. These planes subdivide the model into curved slices, that can slide with respect to each other.

The sliding planes are assumed to be smooth, in order that volume remains unchanged during sliding, and the model therefore represents the situation at progressive flow. The sliding motion is restricted to a relative displacement of adjacent slices in such a direction, that the shear stresses on the planes (see figure 8) dissipate energy. This restriction is due to the friction character of the behaviour of granular media and will be indicated in the following text by the thermodynamic requirement of energy dissipation.

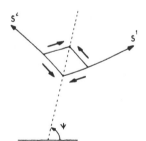

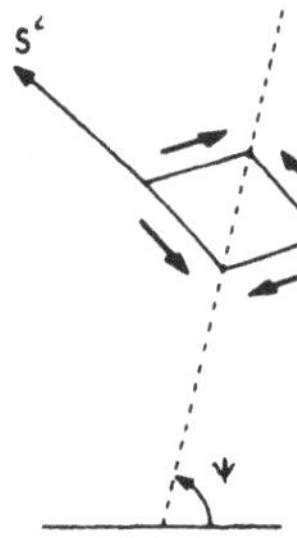

Fig. 8 – Shear stress directions on element faces.

No other restrictions are imposed on the motions of the slices in the model. Slidings are left free to adopt any magnitude. In the two conjugate stress characteristic directions slidings can occur simultaneously and can possess unequal values. Further it is assumed, that the rotation of the slices is free to adopt any magnitude desired. The slice rotation is called *structural* rotation here, because it is the rotation of the grain structure. As shown in section 2.7, this rotation differs from the asymmetric part of the velocity gradient tensor, which generally is called the material rotation, because it represents the rotation of the grain material as a total, when interpreted as a deformable body from the outside.

In order to emphasize the freedoms, attributed to this model on the basis of physical considerations, the name "double sliding, free rotating" was chosen to indicate the model. This name was adopted to distinguish the model from other models proposed, which lack these freedoms and therefore represent granular assemblies, that possess irrealistic internal resistances.

Reducing the freedom of motion has mathematical advantages. For instance requiring, that slidings in the two conjugate directions are equal, prohibits the occurence of noncoaxiality (described in section 2.8.) and leads to coaxial models. Restricting the structural rotation in some manner has the mathematical advantage of reducing the constitutive relations to two equalities. Boundary value problems then have unique solutions, in cases where the free rotating model has a bounded range of solutions (see section 2.9.), because of the occurence of inequalities in the constitutive relations.

The opinion of this author is that overestimation of internal resistance in a model should be avoided and cannot be justified by mathematical convenience. Not mathematics, but physical reality should be basic to model propositions. Therefore, it is of interest that noncoaxiality and freedom of rotation, inherent in the model proposed here, have been observed in tests on photoelastic granular assemblies (Drescher, 1976).

In order to describe sliding of the slices mathematically the discontinuous motion, encountered when blocks of finite size slide with respect to each other along discrete separation planes, is replaced by continuous shear strain rates. Mathematically the slices are replaced by sheets of infinitesimal thickness, which intersect the $x ; y$ plane along lines in s^1, s^2 directions. Scalar quantities $a ; b$ are introduced, whose values indicate the strength of sliding along $s^1 ; s^2$ lines respectively defined by:

a = relative velocity created by the sliding of two adjacent s^1-lines divided by their mutual distance.

b = relative velocity created by the sliding of two adjacent s^2-lines divided by their mutual distance.

The quantities a;b are defined to be positive, when sliding is as in figure 9. Since the thermodynamic requirement of energy dissipation must be satisfied, the scalar quantities a;b are restricted to positive value or

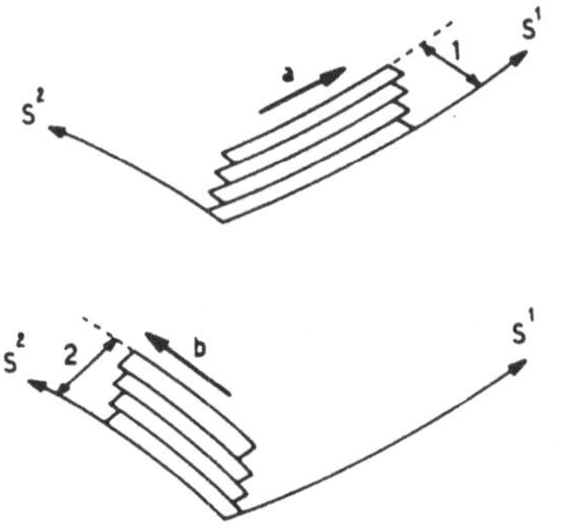

Fig. 9 – Shear strain rates.

$$0 \leq a \leq +\infty; \quad 0 \leq b \leq +\infty. \tag{2.1.1.}$$

The double sliding mechanism admits unequal sliding rates in conjugate directions or

$$a \neq b. \tag{2.1.2.}$$

Besides sliding, the model permits slices to rotate. This rotation is called structural rotation here, and is indicated by a scalar quantity Ω, positive for anticlockwise rotation (figure 10). In the analysis two quantities $\Omega_1;\Omega_2$ are considered, defined by

Ω_1 = structural rotation of a line in s^1-direction.
Ω_2 = structural rotation of a line in s^2-direction.

In order to guarantee, that deformation of the system is created exclusively by slidings, the slices of the model are assumed to conserve inside angles. Mathematically this is obtained by requiring, that the structural rotations of two conjugate $s^1;s^2$-lines in a point are equal to the same value Ω, or

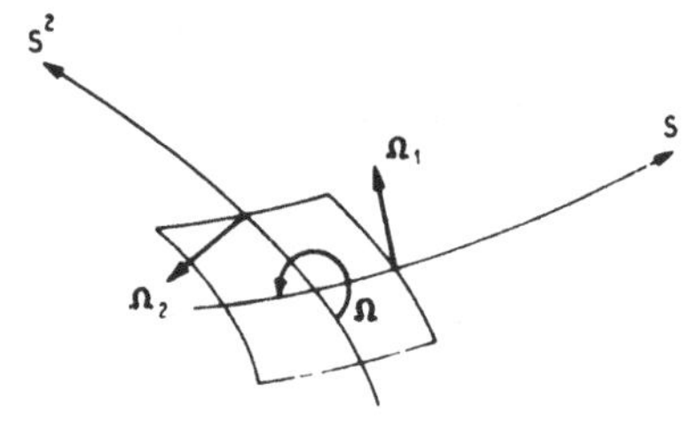

Fig. 10 – Structural rotation of element.

$$\Omega_1 = \Omega_2 = \Omega. \tag{2.1.3.}$$

That the structural rotation is not restricted physically by the conditions at a point, can be written as

$$-\infty \leq \Omega \leq +\infty. \tag{2.1.4.}$$

2.2. a,b and Ω expressed in velocity components of s^1, s^2-system

Since the flow mechanism of the model consists of slidings in the directions of the stress characteristics $s^1; s^2$, it is appropriate to construct mathematical expressions in terms of velocity components in the $s^1; s^2$- coordinate system (figure 11). These components are indicated by a bar over the symbol V for velocity.

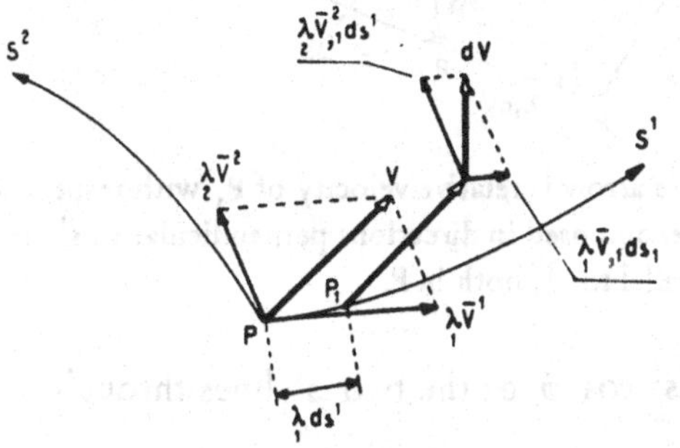

Fig. 11 – Velocity of a point P (white arrow) decomposed in contravariant components of physical magnitude

Using the scale parameters $\underset{1}{\lambda}; \underset{2}{\lambda}$ introduced in section 1.2, the quantities $\underset{1}{\lambda}\bar{V}^1 ; \underset{2}{\lambda}\bar{V}^2$ represent the physical magnitudes of the contravariant velocity components. Let two adjacent points P and P_1 on the same s^1-line be considered, at a mutual infinitesimal distance $\underset{1}{\lambda} ds^1$. Their relative r... ..on, dV, has contravariant components of physical magnitude given by

$$\underset{1}{\lambda}\, \bar{V}^1_{,1}\, ds^1 \qquad ; \qquad \cdot\; \bar{V}^2_{,1}\, ds^1 .$$

The covariant derivatives in these expressions are related to the partial derivatives by

$$\bar{V}^p_{,q} = (\partial \bar{V}^p / \partial s^q) + \Gamma^p_{qr}\, \bar{V}^r \qquad (2.2.1.)$$

where Γ^p_{qr} are Christoffel symbols given in section 2.3.

In the model an s^1-line is not deformed by slidings in s^1-direction. Therefore, only shear strain rate b created by slidings in s^2-direction and a rotation Ω_1 of the s^1-line, contribute to the relative velocity of P_1 with respect to P. In figure 12 the vector dV_b is the contribution of b and dV_p the contribution of Ω_1. The vector dV_b is in the s^2-direction and has a magnitude equal to b multiplied by the

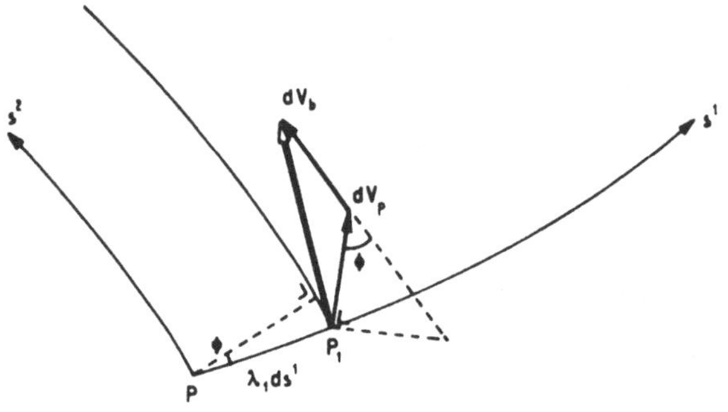

Fig. 12 – White arrow is relative velocity of P_1 with respect to P decomposed in directions perpendicular to s^1 and parallel to s^2, both in P.

perpendicular distance $\underset{1}{\lambda} ds^1 \cos\phi$ of the two s^1-lines through P and P_1. So

$$dV_b = b\cos\phi \, \underset{1}{\lambda}\, ds^1 \, .$$

The vector dV_p is perpendicular to s^1 and has a magnitude

$$dV_p = \Omega_1 \, \underset{1}{\lambda} ds^1 \, .$$

Decomposition of these contributions in s^1 ; s^2 -directions (see figure 12) gives

$$\underset{1}{\lambda}\, \bar{V}^1_{,1}\, ds^1 = dV_p \tan\phi \tag{2.2.2.}$$

$$\underset{2}{\lambda}\, \bar{V}^2_{,1}\, ds^1 = dV_b + dV_p/\cos\phi \tag{2.2.3.}$$

and substitution of dV_p ; dV_b leads to

$$\underset{1}{\lambda}\bar{V}^1_{,1} = \underset{1}{\lambda}\,\Omega_1 \tan\phi \tag{2.2.4.}$$

$$\underset{2}{\lambda}\bar{V}^2_{,1} = \underset{1}{\lambda} b\cos\phi + (\underset{1}{\lambda}\Omega_1/\cos\phi) \, . \tag{2.2.5.}$$

A similar reasoning applied to the relative velocity of an adjacent point P_2 on the s^2-line through P gives

$$\underset{1}{\lambda}\bar{V}^1_{,2} = \underset{2}{\lambda} a \cos\phi - (\underset{2}{\lambda}\Omega_2 / \cos\phi) \tag{2.2.6.}$$

$$\underset{2}{\lambda}\bar{V}^2_{,2} = - \underset{2}{\lambda}\Omega_2 \tan\phi \, . \tag{2.2.7.}$$

Using (2.1.3.) in (2.2.4.) and (2.2.7.) results in

$$\Omega_1 \tan\phi = \Omega_2 \tan\phi = \Omega \tan\phi = \bar{V}^1_{,1} = - \bar{V}^2_{,2} \tag{2.2.8.}$$

which can be reduced to

$$\bar{V}^1_{,1} + \bar{V}^2_{,2} = 0 \, . \tag{2.2.9.}$$

The relation above expresses, that the divergence of the velocity vanishes and this has the physical meaning, that volume is conserved in the model.

The equations (2.2.4.) through (2.2.9.) together with the restrictions (2.1.1.) (2.1.4.) on a ; b and Ω provided a mathematical description of the double sliding, free rotating model.

2.3. Scale parameters, Metric tensor and Christoffel symbols for the s^1, s^2-system

The curvilinear $s^1; s^2$-system is related to the cartesian $x;y$ system by the Jacobi matrix (1.2.4.), whose components repeated here are

$$(2.3.1.)\quad \begin{cases} \dfrac{\partial x}{\partial s^1} = \underset{1}{\lambda} \cos\alpha_1 & \dfrac{\partial y}{\partial s^1} = \underset{1}{\lambda} \sin\alpha_1 \\[2ex] \dfrac{\partial x}{\partial s^2} = \underset{2}{\lambda} \cos\alpha_2 & \dfrac{\partial y}{\partial s^2} = \underset{2}{\lambda} \sin\alpha_2 \,. \end{cases}$$

The inverse can be shown to have the components

$$(2.3.2.)\quad \begin{cases} \dfrac{\partial s^1}{\partial x} = \dfrac{\sin\alpha_2}{\underset{1}{\lambda}\cos\phi} & \dfrac{\partial s^2}{\partial x} = \dfrac{-\sin\alpha_1}{\underset{2}{\lambda}\cos\phi} \\[2ex] \dfrac{\partial s^1}{\partial y} = \dfrac{-\cos\alpha_2}{\underset{1}{\lambda}\cos\phi} & \dfrac{\partial s^2}{\partial y} = \dfrac{\cos\alpha_1}{\underset{2}{\lambda}\cos\phi} \,. \end{cases}$$

The scale parameters $\underset{1}{\lambda}; \underset{2}{\lambda}$ vary as a consequence of the field curvature. Relations describing this dependence can be obtained by satisfying the integrability conditions

$$(2.3.3.)\quad \frac{\partial^2 x}{\partial s^1 \partial s^2} = \frac{\partial^2 x}{\partial s^2 \partial s^1} \quad \text{and} \quad \frac{\partial^2 y}{\partial s^1 \partial s^2} = \frac{\partial^2 y}{\partial s^2 \partial s^1} \,.$$

Elaboration with (2.3.1.) gives relations containing $\partial\underset{1}{\lambda}/\partial s^2 ; \partial\underset{2}{\lambda}/\partial s^1 ; \partial\psi/\partial s^1 ; \partial\psi/\partial s^2$ because in the expressions (1.2.1.) for $\alpha_1 ; \alpha_2$ the angle ψ is a variable. After solving for the derivatives of $\underset{1}{\lambda}; \underset{2}{\lambda}$ there results

$$(2.3.4.)\quad \frac{1}{\underset{1}{\lambda}\underset{2}{\lambda}} \frac{\partial \underset{1}{\lambda}}{\partial s^2} = - \frac{1}{\underset{1}{\lambda}\cos\phi} \frac{\partial\psi}{\partial s^1} - \frac{\sin\phi}{\underset{2}{\lambda}\cos\phi} \frac{\partial\psi}{\partial s^2}$$

$$(2.3.5.)\quad \frac{1}{\underset{1}{\lambda}\underset{2}{\lambda}} \frac{\partial \underset{2}{\lambda}}{\partial s^1} = + \frac{\sin\phi}{\underset{1}{\lambda}\cos\phi} \frac{\partial\psi}{\partial s^1} + \frac{1}{\underset{2}{\lambda}\cos\phi} \frac{\partial\psi}{\partial s^2} \,.$$

The components of the metric tensor are obtained by the rules

$$g_{pq} = \frac{\partial x^\alpha}{\partial s^p}\frac{\partial x^\alpha}{\partial s^q} \quad ; \quad g^{pq} = \frac{\partial s^p}{\partial x^\alpha}\frac{\partial s^q}{\partial x^\alpha}$$

where $x^1 = x$, $x^2 = y$ and the Einstein summation convention is implied. This gives with (2.3.1.) and (2.3.2.)

$$\left.\begin{array}{lll} g_{11} = \lambda_1^2 & g_{12} = g_{21} = -\lambda_1\lambda_2 \sin\phi & g_{22} = \lambda_2^2 \\ g^{11} = 1/\lambda_1^2\cos^2\phi & g^{12} = g^{21} = \sin\phi/\lambda_1\lambda_2\cos^2\phi & g^{22} = 1/\lambda_2^2\cos^2\phi \end{array}\right\} \quad (2.3.6.)$$

the Christoffel symbols Γ^p_{qr} are defined by

$$\Gamma^p_{qr} = \frac{\partial s^p}{\partial x^\alpha}\,\frac{\partial^2 x^\alpha}{\partial s^q\,\partial s^r} \,. \qquad (2.3.7.)$$

After elaboration with the formulas in this section, these can be shown to be

$$\left.\begin{array}{ll} \Gamma^1_{11} = \dfrac{1}{\lambda_1}\dfrac{\partial\lambda_1}{\partial s^1} + \tan\phi\,\dfrac{\partial\psi}{\partial s^1} & \Gamma^2_{11} = +\dfrac{\lambda_1}{\lambda_2\cos\phi}\dfrac{\partial\psi}{\partial s^1} \\ \Gamma^1_{12} = \Gamma^1_{21} = -\dfrac{\lambda_2}{\lambda_1\cos\phi}\dfrac{\partial\psi}{\partial s^1} & \Gamma^2_{12} \quad \Gamma^2_{21} = +\dfrac{\lambda_1}{\lambda_2\cos\phi}\dfrac{\partial\psi}{\partial s^2} \\ \Gamma^1_{22} = -\dfrac{\lambda_2}{\lambda_1\cos\phi}\dfrac{\partial\psi}{\partial s^2} & \Gamma^2_{22} = \dfrac{1}{\lambda_2}\dfrac{\partial\lambda_2}{\partial s^2} - \tan\phi\,\dfrac{\partial\psi}{\partial s^2} \,. \end{array}\right\} \quad (2.3.8.)$$

2.4. Compatibility expressed in derivatives of a,b and Ω.

The appellation "free rotating" was added to the model name in order to emphasize, that the structural rotation is not restricted by the conditions at a point. The structural rotation of the slices is free to adopt any value needed for a slice to adjust itself to the motions of adjacent slices. Because this adjustment is necessary, the magnitudes of the derivatives of Ω are not free, but restricted by requirements of compatibility. Formulated mathematically in terms of the physical components of velocity in the $s^1 ; s^2$ system, the compatibility requirements are

$$(2.4.1.)\qquad \frac{\partial^2(\underset{1}{\lambda}\bar{V}^1)}{\partial s^1\,\partial s^2} = \frac{\partial^2(\underset{1}{\lambda}\bar{V}^1)}{\partial s^2\,\partial s^1} \quad ; \quad \frac{\partial^2(\underset{2}{\lambda}\bar{V}^2)}{\partial s^1\,\partial s^2} = \frac{\partial^2(\underset{2}{\lambda}V^2)}{\partial s^2\,\partial s^1}\,.$$

These requirements can be expressed in terms of Ω, a and b by eliminating the velocity components with relations (2.2.4.) through (2.2.7.). In order to apply these relations it is necessary to transform the covariant derivatives of the velocity components into partial derivatives. This can be done by use of the rule (2.2.1.). For example relation (2.2.4.) then becomes with (2.3.8.)

$$(2.4.2.)\qquad \begin{aligned} \underset{1}{\lambda}\Omega\tan\phi &= \underset{1}{\lambda}\bar{V}^1_{,1} = \underset{1}{\lambda}(\partial\bar{V}^1/\partial s^1) + \underset{1}{\lambda}\Gamma^1_{1r}\bar{V}^r = \\ &= \underset{1}{\lambda}\frac{\partial\bar{V}^1}{\partial s^1} + \frac{\partial\underset{1}{\lambda}}{\partial s^1}\bar{V}^1 + \underset{1}{\lambda}\tan\phi\,\frac{\partial\psi}{\partial s^1}\bar{V}^1 - \frac{\underset{2}{\lambda}}{\cos\phi}\,\frac{\partial\psi}{\partial s^1}\bar{V}^2. \end{aligned}$$

A similar elaboration of (2.2.4.) through (2.2.7.) and use of (2.3.4.) (2.3.5.) gives the following set of relations

$$(2.4.3.)\qquad \underset{1}{\lambda}\,\Omega\tan\phi = \frac{\partial(\underset{1}{\lambda}\bar{V}^1)}{\partial s^1} + \frac{1}{\cos\phi}\,\frac{\partial\psi}{\partial s^1}\,(\underset{1}{\lambda}\sin\phi\,\bar{V}^1 - \underset{2}{\lambda}\bar{V}^2)$$

$$(2.4.4.)\qquad \underset{1}{\lambda}\,b\cos\phi + \underset{1}{\lambda}\Omega\,\frac{1}{\cos\phi} = \frac{\partial(\underset{2}{\lambda}\bar{V}^2)}{\partial s^2} + \frac{1}{\cos\phi}\,\frac{\partial\psi}{\partial s^1}\,(\underset{1}{\lambda}\bar{V}^1 - \underset{2}{\lambda}\sin\phi\,\bar{V}^2)$$

$$(2.4.5.)\qquad \underset{2}{\lambda}\,a\cos\phi \quad \underset{2}{\lambda}\Omega\,\frac{1}{\cos\phi} = \frac{\partial(\underset{1}{\lambda}\bar{V}^1)}{\partial s^2} + \frac{1}{\cos\phi}\,\frac{\partial\psi}{\partial s^2}\,(\underset{1}{\lambda}\sin\phi\,\bar{V}^1 - \underset{2}{\lambda}\bar{V}^2)$$

$$(2.4.6.)\qquad \underset{2}{\lambda}\,\Omega\tan\phi = \frac{\partial(\underset{2}{\lambda}\bar{V}^2)}{\partial s^2} + \frac{1}{\cos\phi}\,\frac{\partial\psi}{\partial s^2}\,(\underset{1}{\lambda}\bar{V}^1 - \underset{2}{\lambda}\sin\phi\,\bar{V}^2)\,.$$

By elaboration of the combination

$$- \frac{\partial(2.4.3)}{\partial s^2} + \sin\phi \frac{\partial(2.4.4)}{\partial s^2} + \frac{\partial(2.4.5)}{\partial s^1} - \sin\phi \frac{\partial(2.4.6)}{\partial s^1}$$

it is found that

$$\sin\phi\cos\phi \frac{\partial(\underset{1}{\lambda} b)}{\partial s^2} - \cos\phi \frac{\partial(\underset{2}{\lambda}\Omega)}{\partial s^1} + \cos\phi \frac{\partial(\underset{2}{\lambda} a)}{\partial s^1} = \text{with (2.4.1)}$$

$$= + \frac{1}{\cos\phi} \frac{\partial}{\partial s^2} [\frac{\partial\psi}{\partial s^1} (\cos^2\phi \underset{2}{\lambda} \bar{V}^2)] + \frac{1}{\cos\phi} \frac{\partial}{\partial s^1} [\frac{\partial\psi}{\partial s^2} (-\cos^2\phi \underset{2}{\lambda} \bar{V}^2)] =$$

$$= \cos\phi [\frac{\partial\psi}{\partial s^1} \frac{\partial(\underset{2}{\lambda} \bar{V}^2)}{\partial s^2} - \frac{\partial\psi}{\partial s^2} \frac{\partial(\underset{2}{\lambda} \bar{V}^2)}{\partial s^1}] = \text{with (2.4.4) and (2.4.6)}$$

$$= \cos\phi [- \underset{2}{\lambda}\Omega \tan\phi \frac{\partial\psi}{\partial s^1} - \underset{1}{\lambda} b \cos\phi \frac{\partial\psi}{\partial s^2} - \underset{1}{\lambda}\Omega \frac{1}{\cos\phi} \frac{\partial\psi}{\partial s^2}] = \text{with (2.3.5)}$$

$$= - \cos\phi\, \Omega \frac{\partial\underset{2}{\lambda}}{\partial s^1} - \underset{1}{\lambda} b \cos^2\phi \frac{\partial\psi}{\partial s^2} .$$

This can be written as

$$\sin\phi \frac{\partial(\underset{1}{\lambda} b)}{\partial s^2} + \frac{\partial(\underset{2}{\lambda} a)}{\partial s^1} = - (\underset{1}{\lambda} b) \cos\phi \frac{\partial\psi}{\partial s^2} + \underset{2}{\lambda} \frac{\partial\Omega}{\partial s^1} . \qquad (2.4.6)$$

By a similar elaboration of

$$\sin\phi \frac{\partial(2.4.3)}{\partial s^2} - \frac{\partial(2.4.4)}{\partial s^2} + \sin\phi \frac{\partial(2.4.5)}{\partial s^1} - \frac{\partial(2.4.6)}{\partial s^1}$$

it is found that

$$\sin\phi \frac{\partial(\underset{2}{\lambda} a)}{\partial s^1} + \frac{\partial(\underset{1}{\lambda} b)}{\partial s^2} = + (\underset{2}{\lambda} a) \cos\phi \frac{\partial\psi}{\partial s^1} - \underset{1}{\lambda} \frac{\partial\Omega}{\partial s^2} . \qquad (2.4.7)$$

Since in the relations (2.4.6.) and (2.4.7.) terms appear that can be interpreted as the magnitude of bending of the elements, we will call these the *bending relations.*

2.5. Character of the bending relations

The various terms in (2.4.6.) can be interpreted physically as follows. The quantity $(\partial\psi/\partial s^2)$ is the curvature of an s^2-line. When the curvature is as in figure 13, this quantity is positive, because ψ is the angle of the bissectrix of the equiangular s^1; s^2 -system and the x-axis. Because of (2.1.1.) b must be positive. Since also the scale parameter λ_1 is a positive quantity, the entire term $(\lambda_1 b) \cos\phi(\partial\psi/\partial s^2)$ is positive.

A shear strain rate b active on curved slices, produces an increase of structural rotation, when subsequent slices in s^1-direction are considered. The quantity $(\partial\Omega/\partial s^1)$ is positive, when there is an increase of structural rotation in s^1-direction. If the two last terms in (2.4.6.) are equal, such that

$$(\lambda_1 b) \cos\phi \ (\partial\psi/\partial s^2) = \lambda_2(\partial\Omega/\partial s^1) \qquad (2.5.1.)$$

the structural-rotation is exclusively due to the curvature of the slices and the slices can remain internally rigid during the motion.

When, however, the terms differ, such that

$$(\lambda_1 b) \cos\phi \ (\partial\psi/\partial s^2) \neq \lambda_2(\partial\Omega/\partial s^1) \qquad (2.5.2.)$$

the rotation variation along the s^1-line can only be followed by the slice, if there is internal deformation in the form of a bending rate. (see figure 14).

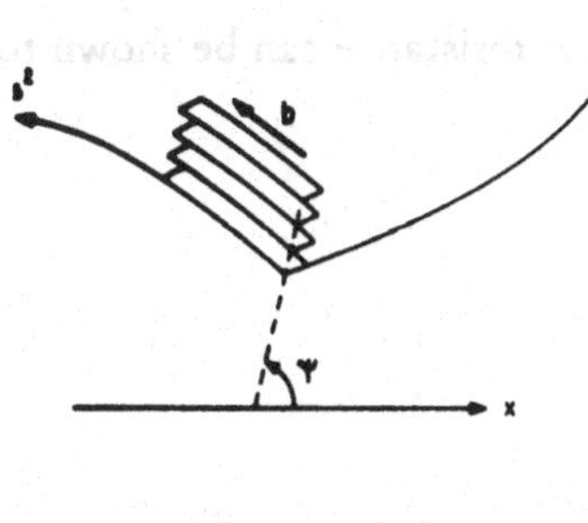

Fig. 13 – Curvature of s^2-lines corresponding to positive value of $\partial\psi/\partial s^2$.

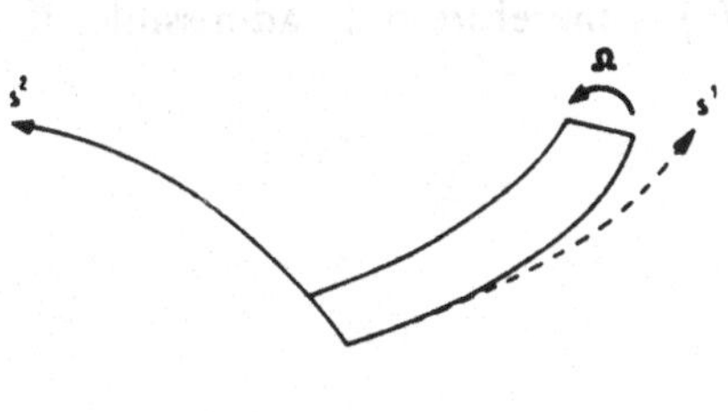

Fig. 14 – Bending of a slice, when $\partial\Omega/\partial s^1$ is positive.

For solving boundary value problems it would be convenient to propose (2.5.1.) to be valid, because the relation (2.4.6.) then reduces to

(2.5.3.) $$\sin\phi\,[\,\partial(\underset{1}{\lambda}b)\,/\,\partial s^2\,] + [\,\partial(\underset{2}{\lambda}a)\,/\,\partial s^1\,] = 0\ .$$

Similarly, if it could be stated that

(2.5.4.) $$(\underset{2}{\lambda}a)\,\cos\phi\,(\partial\psi/\partial s^1) = \underset{1}{\lambda}(\partial\Omega\,/\,\partial s^2)$$

then relation (2.4.7.) would reduce to

(2.5.5.) $$\sin\phi\,[\,\partial(\underset{2}{\lambda}a)\,/\,\partial s^1\,] + [\,\partial(\underset{1}{\lambda}b)\,/\,\partial s^2\,] = 0$$

and the relations (2.5.3.) (2.5.5.) would give

(2.5.6.) $$\partial(\underset{1}{\lambda}a)\,/\,\partial s^1 = 0 \quad ; \quad \partial(\underset{2}{\lambda}b)\,/\,\partial s^2 = 0\ .$$

The convenience is, that the relations (2.5.6.) can be integrated and would yield an unique solution for certain boundary value problems. Declaring (2.5.6.) valid, however, includes the danger that the model is attributed a resistance against bending, which a granular assembly represented by the model may not possess. The restrictions on Ω imposed by (2.5.1.) (2.5.4.) may correspond to an overestimation of the capacity of grain agglomerates to resist internal deformation. Application of (2.5.6.) is therefore only admissible, if the internal resistance can be shown to exist.

2.6. Constitutive relations

The equations (2.2.4.) through (2.2.7.) describe the flow by relating the velocity components to the shear strain rates a;b and the structural rotation Ω . In order to obtain constitutive relations from these, it is necessary to introduce the properties of the scalars a;b; Ω derived from the physical considerations in section 2.1.

It is convenient to convert the relevant relations first in cartesian coordinates $x^1 = x$, $x^2 = y$. Velocity components in these coordinates will be indicated by V without overbar. Because the system is cartesian, there is no difference between contravariant and covariant components, so

$$V^{\alpha} = V_{\alpha} .$$

The transformation rules for the transition from s^1, s^2 to x^1, x^2 coordinates are

$$\bar{V}^p = V^{\alpha} \frac{\partial s^p}{\partial x^{\alpha}} = V_{\alpha} \frac{\partial s^p}{\partial x^{\alpha}}$$

$$\bar{V}^p_{,q} = V^{\alpha}_{,\beta} \frac{\partial s^p}{\partial x^{\alpha}} \frac{\partial x^{\beta}}{\partial s^q} = V_{\alpha,\beta} \frac{\partial s^p}{\partial x^{\alpha}} \frac{\partial x^{\beta}}{\partial s^q} .$$

Solving for b; Ω_1 from (2.2.4.) (2.2.5.) and for a; Ω_2 from (2.2.6.) (2.2.7.) and applying the transformation rules gives

$$a \sin\phi \cos\phi = - V_{x,x} \cos^2\alpha_2 - (V_{x,y} + V_{y,x}) \cos\alpha_2 \sin\alpha_2 - V_{y,y} \sin^2\alpha_2 \qquad (2.6.1.)$$

$$b \sin\phi \cos\phi = - V_{x,x} \cos^2\alpha_1 - (V_{x,y} + V_{y,x}) \cos\alpha_1 \sin\alpha_1 - V_{y,y} \sin^2\alpha_1 \qquad (2.6.2.)$$

$$\Omega_1 \sin\phi = + V_{x,x} \cos\alpha_1 \sin\alpha_2 + V_{x,y} \sin\alpha_1 \sin\alpha_2 - V_{y,x} \cos\alpha_1 \cos\alpha_2 - V_{y,y} \sin\alpha_1 \cos\alpha_2 \qquad (2.6.3.)$$

$$\Omega_2 \sin\phi = + V_{x,x} \cos\alpha_2 \sin\alpha_1 + V_{x,y} \sin\alpha_2 \sin\alpha_1 - V_{y,x} \cos\alpha_2 \cos\alpha_1 - V_{y,y} \sin\alpha_2 \cos\alpha_1 . \qquad (2.6.4.)$$

Applying the physical requirement of conservation of inside angles (2.2.8.) it is obtained with (1.2.1.) that

$$V_{x,x} + V_{y,y} = 0 \tag{2.6.5.}$$

and

$$\varphi = (V_{x,x} - V_{y,y}) \sin 2\psi - (V_{x,y} + V_{y,x}) \cos 2\psi + + (- V_{x,y} + V_{y,x}) \sin \phi \, . \tag{2.6.6.}$$

Relation (2.6.5.) is identical to (2.2.9.) and indicates, that the model possesses the property to conserve volume.

Using (1.2.1.) (2.6.5.) in (2.6.1.) (2.6.2.) results in

$$a \sin 2\phi = + (V_{x,x} - V_{y,y}) \sin (2\psi + \phi) - (V_{x,y} + V_{y,x}) \cos (2\psi + \phi) \tag{2.6.7.}$$

$$b \sin 2\phi = - (V_{x,x} - V_{y,y}) \sin (2\psi - \phi) + (V_{x,y} + V_{y,x}) \cos (2\psi - \phi) . \tag{2.6.8.}$$

In section 2.1. the physical properties of the shear strain rates a,b and the structural rotation Ω are expressed by (2.1.1.) (2.1.4.), which repeated here are

$$0 \leqslant a \leqslant + \infty \tag{2.6.9.}$$

$$0 \leqslant b \leqslant + \infty \tag{2.6.10.}$$

$$- \infty \leqslant \Omega \leqslant + \infty \, . \tag{2.6.11.}$$

Together with the relations (2.6.5.) through (2.6.8.) they form the constitutive relations available for establishing velocity distributions from boundary conditions.

Since Ω is not restricted in any manner, as indicated by (2.6.11.), equation (2.6.6.) cannot be used for solving boundary value problems. So there remain the three relations (2.6.5.) (2.6.7.) (2.6.8.). They form an unusual set, because the limitations (2.6.9.) (2.6.10.) only specify that a,b must be positive. Since the angle of internal friction is restricted to the range

$$0 \leqslant \phi < \frac{1}{2}\pi \tag{2.6.12}$$

also sin 2ϕ is positive. So the set of constitutive relations has the following form

$$V_{x,x} + V_{y,y} = 0 \qquad (2.6.13.)$$

$$+ (V_{x,x} - V_{y,y}) \sin(2\psi + \phi) - (V_{x,y} + V_{y,x}) \cos(2\psi + \phi) > 0 \qquad (2.6.14.)$$

$$- (V_{x,x} - V_{y,y}) \sin(2\psi - \phi) + (V_{x,y} + V_{x,y}) \cos(2\psi - \phi) > 0\,. \qquad (2.6.15)$$

This set of constitutive relations has the unusual character of containing inequalities. This has the inconvenient consequence, that boundary value problems have a range of solutions, instead of one unique solution usually obtained from a system consisting of equalities. Other models have been proposed for the sliding mechanism that do consist of equalities, by restricting a;b and/or Ω in some manner. Since these restrictions have no physical basis, introduction in the model leads to an overestimation of the resistance of the represented granular assembly.

2.7. Structural rotation Ω and asymmetric part of velocity gradient tensor ω.

By its definition in section 2.1., the quantity Ω is the structural rotation of the sliding slices. This is a different quantity from ω defined by

$$2\omega = -\ 2\omega_{yx} = 2\omega_{xy} = (V_{y,x} - V_{x,y}) \tag{2.7.1.}$$

which is the asymmetric part of the velocity gradient tensor. The quantity ω is the rotation of a body, as observed from the outside. A relation between Ω and ω can be obtained as follows.

From (2.6.7.) (2.6.8.) it follows that

$$(a - b)\sin\phi = (V_{x,x} - V_{y,y})\sin 2\psi - (V_{x,y} + V_{y,x})\cos 2\psi\ . \tag{2.7.2}$$

Introduced in (2.6.6.) this gives with (2.7.1.)

$$2\Omega - a + b = 2\omega\ . \tag{2.7.3}$$

This relation shows that the apprent rotation, ω, of the system as observed from the outside, consists of structural rotation Ω of the slices minus the difference of the shear strain rates, a and b.

2.8. Non coaxiality, and lack of normality

The strain rate is a tensor with components $\dot{\epsilon}_{ij}$, that are related to the components of the velocity gradient tensor $V_{i,j}$ by

$$\dot{\epsilon}_{xx} = V_{x,x} \quad ; \quad \dot{\epsilon}_{xy} = \dot{\epsilon}_{yx} = \frac{1}{2}(V_{yx} + V_{y,x}) \quad ; \quad \dot{\epsilon}_{yy} = V_{y,y} \,. \tag{2.8.1}$$

The sliding slices model is rigid-plastic. Therefore the total strain rates are eq .ai 'o the plastic strain rates.

Let twice the strain rate deviator W be introduced as a positive quantity, equal to

$$W = [(\dot{\epsilon}_{xx} - \dot{\epsilon}_{yy})^2 + (\dot{\epsilon}_{xy} + \dot{\epsilon}_{yx})^2]^{1/2}$$

and let θ be the angle between the x-axis and the major principal strain rate. Then

$$\begin{aligned} \dot{\epsilon}_{xx} - \dot{\epsilon}_{yy} &= V_{x,x} - V_{y,y} = W\cos 2\theta \\ \dot{\epsilon}_{xy} + \dot{\epsilon}_{yx} &= V_{x,y} + V_{y,x} = W\sin 2\theta \end{aligned} \tag{2.8.2}$$

Used in (2.6.7.) (2.6.8.) there results

$$a \sin 2\phi = - W \sin(2\theta - 2\psi - \phi) \geqslant 0 \tag{2.8.3}$$

$$b \sin 2\phi = + W \sin(2\theta - 2\psi + \phi) \geqslant 0 \tag{2.8.4}$$

$$(a + b) \cos\phi = + W \cos(2\theta - 2\psi) \geqslant 0 \tag{2.8.5}$$

$$(a - b) \sin\phi = - W \sin(2\theta - 2\psi) \,. \tag{2.8.6}$$

The three expressions (2.8.3) (2.8.4.) (2.8.5.) must be positive definite, in order to satisfy the thermodynamic requirement of energy dissipation expressed by (2.1.1.) and the restriction (2.6.12.) imposed on ϕ .

The inequalities (2.8.3.) and (2.8.4.) can be written as a continued inequality

$$- W\cos(2\theta - 2\psi)\sin\phi \leqslant W\sin(2\theta - 2\psi)\cos\phi \leqslant W\cos(2\theta - 2\psi)\sin\phi \,. \tag{2.8.7}$$

Because of (2.8.5.) and (2.6.12.), W $\cos(2\theta - 2\psi)$ and $\cos\phi$ are positive definite and (2.8.7.) can therefore be divided by those terms to give

$$(2.8.8) \qquad -\tan\phi \leqslant \tan(2\theta - 2\psi) \leqslant \tan\phi \; .$$

Introduce i as the deviation angle between the directions of major principal stress and major principal strain rate, such that

$$(2.8.9) \qquad i = \theta - \psi \, ,$$

then from (2.8.8.) it follows, that i is restricted by

$$(2.8.10) \qquad -\frac{1}{2}\phi \leqslant i \leqslant +\frac{1}{2}\phi \; .$$

The possibility for i to differ from zero, indicates that in the model the principal directions of stress and strain rate can deviate. This deviation is called *noncoaxiality*.

The magnitude of i cannot be established a priori. It is not a material property, that can be determined from tests. That its value can differ from zero, is a consequence of the possibility, that a can differ from b. This can be seen by combining (2.8.5.) and (2.8.6.) to form with (2.8.9.)

$$(2.8.11) \qquad [(b-a)/(b+a)]\tan\phi = \tan 2i \; .$$

This expression shows, that only for the case that a is equal to b the value of i is zero.

It has often been argued, that the noncoaxiality inherent in the double sliding model, is indicative for the existence of an anisotropy. This is a fallacy, originated by applying indiscriminantly a general principle, valid in mechanics. The general rule in question is, that in an isotropic medium the tensors of strain and incremental stress are coaxial, if there exists an unique functional relationship between the two. Application cf this rule is not allowed in this case, because the rule is no more valid, when a material is in the limit stress state.

When a material is in the limit state of stress, strains can develop although stresses remain constant. Therefore, there is no unique functional relationship between strain and stress increment. As a consequence strains can have any arbitrary magnitude, they can be different in two conjugate directions. So the rule is no more valid and noncoaxiality can develop, although the system is istropic.

The double sliding model considered here is isotropic in the sense, that its behaviour is independent of the principal stress directions. In all directions the same kind of noncoaxial behaviour can occur, because the stresses are in the limit state and a unique functional relationship between strains and stress increments does not exist.

In order to verify whether the double sliding model possesses the property of *normality*, it is necessary to consider the quantity N defined by

$$N = (\sigma_{ij} - \sigma^*_{ij})\,\dot{\epsilon}_{ij} \ . \tag{2.8.12}$$

In this expression σ_{ij} is a limit stress state and $\dot{\epsilon}_{ij}$ is the strainrate occurring under influence of σ_{ij}. Normality exists, when for all stress states σ^*_{ij}, that can be supported by the material, the quantity N is not negative.

For σ_{ij} the relations (1.1.5.) can be taken and for σ^*_{ij} the corresponding forms

$$\left.\begin{aligned} \sigma^*_{xx} &= -\,p^* + p^* \sin\phi^* \cos 2\psi^* + c \operatorname{cotan}\phi \\ \sigma^*_{xy} &= \sigma^*_{yx} = p^* \sin\phi^* \sin 2\psi^* \\ \sigma^*_{yy} &= -\,p^* - p^* \sin\phi^* \cos 2\psi^* + c \operatorname{cotan}\phi \ . \end{aligned}\right\} \tag{2.8.13}$$

In order, that σ^*_{ij} is a possible stress state, the following inequalities must be satisfied

$$p^* \geqslant 0 \quad ; \quad 0 \leqslant \phi^* \leqslant \phi \ .$$

Using (1.1.5.) (2.8.13.) and (2.8.2.) in (2.8.12.) gives

$$N = (-\,p + p^*)(\dot{\epsilon}_{xx} + \dot{\epsilon}_{yy}) +$$

$$+\ W[\,p \sin\phi \cos(2\psi - 2\theta) - p^* \sin\phi^* \cos(2\psi^* - 2\theta)] \ .$$

The first term is zero, because of (2.6.13.) and (2.8.1.). Further applying (2.8.9.) gives

$$N = W[p \sin\phi \cos 2i - p^* \sin\phi^* \cos(2\psi^* - 2\theta)] \,. \tag{2.8.14}$$

The strainrate deviator W is defined as positive and because of (2.8.10.) the first term between brackets satisfies

$$W p \sin\phi \cos 2i \geqslant W p \sin\phi \cos\phi \geqslant 0 \,.$$

However the second term between brackets in (2.8.14.) can be larger than the first term, because p^* can have any positive value. Therefore N can be negative and normality is absent.

The lack of normality is primarily due to the conservation of volume, inherent in the behaviour of granular materials at progressive flow and therefore attributed to the slices model. Conservation of volume is reflected in the equation (2.6.13.), which forms part of the constitutive relations. In the second place, also the noncoaxiality due to the difference in sliding rates, $a \neq b$, contributes to the lack of normality.

There exists no physical law, that requires a material to possess normality. The lack of normality occurring in granular materials violates no basic principles. It is only inconvenient, because the proofs developed for many theorems useful in the theory of plasticity loose their validity. In particular it may be mentioned here that the theorem of uniqueness of stresses in the regions of failure during collapse is no more valid. This implies that the modes of failure depend on the initial stress state.

2.9. Use of constitutive relations in boundary value problems

The constitutive relations were shown in section 2.6. to consist of an equation (2.6.13.) and two inequalities (2.6.14.); (2.6.15.). The equation (2.6.13.) expresses the fact that the model conserves volume. The inequalities (2.6.14.) (2.6.15.) reflect the theormodynamic requirement, that energy is dissipated in both conjugate sliding modes separately.

When used for determining velocity distributions from boundary values, this system gives an answer consisting of a possible range of solutions instead of one unique solution. This is due to the special character of the system, because of the inequalities.

Multiplicators

In order to investigate this special character and establishing the solution procedure, it is useful to introduce multiplicators $\alpha ; \beta$. These are constants independent of $x ; y$ and defined in terms of an auxiliary angle j by

$$\alpha = \cos(2j + \phi) \quad ; \quad \beta = \cos(2j - \phi) \, . \tag{2.9.1}$$

Depending on the value of j, the multiplicators are positive or negative. There are four possible combinations in this respect, corresponding to four regions of j as follows (with k integer):

Interval (1) : $-\frac{1}{4}\pi + \frac{1}{2}\phi < (j + k\pi) < \frac{1}{4}\pi - \frac{1}{2}\phi$, then $\alpha > 0 \; ; \; \beta > 0$.

Interval (2) : $\frac{1}{4}\pi - \frac{1}{2}\phi < (j + k\pi) < \frac{1}{4}\pi + \frac{1}{2}\phi$, then $\alpha < 0 \; ; \; \beta > 0$.

Interval (3) : $\frac{1}{4}\pi + \frac{1}{2}\phi < (j + k\pi) < \frac{3}{4}\pi - \frac{1}{2}\phi$, then $\alpha < 0 \; ; \; \beta < 0$.

Interval (4) : $\frac{3}{4}\pi - \frac{1}{2}\phi < (j + k\pi) < \frac{3}{4}\pi + \frac{1}{2}\phi$, then $\alpha > 0 \; ; \; \beta < 0$.

Specific length increase

The multiplicators $\alpha ; \beta$ are used in c onnection with the shear strain rates $a ; b$ to form a quantity c defined by

$$2c = \alpha a + \beta b \, . \tag{2.9.2.}$$

Since a and b must satisfy the thermodynamic requirement of energy dissipation expressed by (2.1.1.) the value of c in the different internals, is limited as follows

$$\text{(2.9.3)}\qquad \begin{cases} \text{Interval (1)}: & c \geqslant 0 \\ \text{Interval (2)}: & c \text{ is unlimited} \\ \text{Interval (3)}: & c \leqslant 0 \\ \text{Interval (4)}: & c \text{ is unlimited} . \end{cases}$$

Using (2.6.7.) (2.6.8.) in (2.9.2.) gives with (2.9.1.)

$$2c = (V_{x,x} - V_{y,y}) \cos (2\psi + 2j) + (V_{x,y} + V_{y,x}) \sin (2\psi + 2j)$$

Adding (2.6.13.) gives

$$\text{(2.9.4)}\qquad \begin{aligned} c = {} & V_{x,x} \cos^2 (\psi + j) + V_{x,y} \cos (\psi + j) \sin (\psi + j) + \\ & + V_{y,x} \sin (\psi + j) \cos (\psi + j) + V_{y,y} \sin^2 (\psi + j) . \end{aligned}$$

Let orthogonal coordinates $r^1 ; r^2$ be introduced, that deviate by the angle j from the principal stress directions, such that the angle between r^1 and x is $(\psi + j)$ and the angle between r^1 and x is $(\psi + j + 1/2\,\pi)$, see figure 15.

When $\rho_1 ; \rho_2$ are the scale parameters, the Jacobi matrix has coefficients:

$$\text{(2.9.5)}\qquad \begin{cases} (\partial x/\partial r^1) = \rho_1 \cos (\psi + j) & (\partial y/\partial r^1) = \rho_1 \sin (\psi + j) \\ (\partial x/\partial r^2) = - \rho_2 \sin (\psi + j) & (\partial y/\partial r^2) = \rho_2 \cos (\psi + j) \\ \\ \partial r^1/\partial x) = (1/\rho_1) \cos (\psi + j) & (\partial r^2/\partial x) = - (1/\rho_2) \sin (\psi + j) \\ \partial r^1/\partial y) = (1/\rho_1) \sin (\psi + j) & (\partial r^2/\partial y) = (1/\rho_2) \cos (\psi + j) . \end{cases}$$

Then (2.9.4.) can be written as

$$\text{(2.9.6)}\qquad c = V^{\kappa}_{,\lambda} (\partial r^1/\partial x^{\kappa}) (\partial x^{\lambda}/\partial r^1)$$

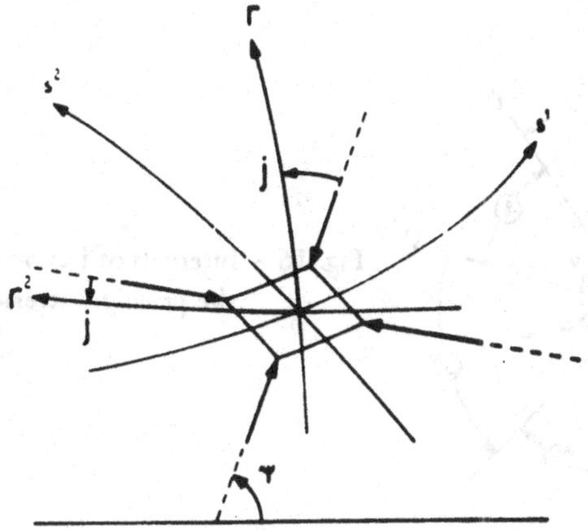

Fig. 15 – Curvilinear coordinate system, r^1 ; r^2.

with V_κ replaced by V^κ, because x ; y is cartesian. Indicating velocity components in the r^1 ; r^2 system by a tilde, this becomes

$$c = \tilde{V}^1_{,1} .$$

The above expression has the physical interpretation, that c represents the specific length increase of a line oriented along r^1.

When the shear strain rates a ; b are both zero, then c vanishes according to (2.9.2.) for all combinations of α ; β . This means for all values of j and therefore all possible orientations of r^1. As a consequence every line in the body, whatever its orientation conserves its length, and this corresponds to a rigid body motion.

Acceptable boundary conditions

When a and/or b are not zero, c can have values different from zero. There are, however, the limitations on c given by (2.9.3.) depending on the interval occupied by the angle j. The respective intervals are shown in figure 16, where j is plotted with respect to the smaller principal compression stress, such that the location of r^1 in the intervals becomes apparent. Since c can be interpreted as the specific length increase of a line oriented along r^1, the limitation described by (2.9.3.) can be specified as follows:

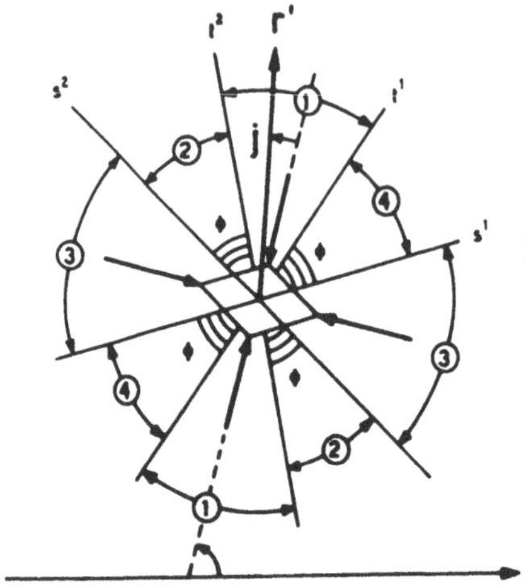

Fig. 16 – Intervals of j as oriented with respect to the principal stresses (small heavy arrows).

When the deviation angle j, between a line and the smaller principal compression stress, is in the interval (1) the line can only elongate, when in interval (3) the line can only shorten, when in the intvals (2) or (4) there is no limitation, the line can either elongate or shorten.

This has a consequence with respect to boundary conditions, consisting of velocity distributions. A boundary at an angle μ can be considered as a line deviating from the plane of major principal compression stress by an angle

$$j = \mu - \psi \, .$$

The interval corresponding to this value of j determines whether the boundary motion is restricted to elongation or shortening. The velocity distribution imposed on the boundary is only acceptable when these restrictions are not violated.

Construction of a velocity distribution

In order to construct velocity distributions from boundary conditions, the relation (2.9.6.) can be used, either for a mathematical integration procedure, or a graphical construction in the hodograph plane.

The *mathematical* elaboration requires to convert the covariant derivative of (2.9.6.) into a partial derivative. This can be done by use of the rule (2.2.1.), which in this case leads to

$$c = \tilde{V}^1_{,1} = (\partial \tilde{V}^1/\partial r^1) + \Gamma^1_{1t}\, \tilde{V}^t. \tag{2.9.7}$$

The Christoffel symbols in this expression can be found by applying the rule (2.3.7.) on the coefficients (2.9.5.), giving

$$\Gamma^1_{11} = (1/\rho_1)(\partial \rho_1/\partial r^1)$$

$$\Gamma^1_{12} = -(\rho_2/\rho_1)\,[\partial(\psi + j)/\partial r^1]\ .$$

So (2.9.7.) becomes

$$c = \frac{\partial(\rho_1 \tilde{V}^1)}{\rho_1\, \partial r^1} - (\rho_2\, \tilde{V}^2)\, \frac{\partial(\psi + j)}{\rho_1\, \partial r^1}\ . \tag{2.9.8}$$

In this expression $\rho_1 \tilde{V}^1 ; \rho_2 \tilde{V}^2$ are the physical velocity components in the $r^1 ; r^2$ coordinate directions. Since further only differentials in r^1 direction are involved, it is possible to integrate (2.9.8.), if the magnitude of c is known. However, only the sign of c is known and, therefore, the solution consists of bounds limiting possible velocity distributions. Representation of these bounds in mathematical terms is somewhat intricate, better suited for comprehension is to express them graphically.

The *graphical* procedure consists in constructing the above mentioned bounds in a hodograph plane. Every point P in the physical x ; y plane has an image P′ in the hodograph plane, such that the coordinates P′ are the velocity components of P. The vector P′Q′ in the hodograph plane is the vector representing the relative velocity of the point Q with respect to the point P (see figure 17).

When Q is located on a r^1-line through P at a distance of magnitude $\rho_1\, dr^1$, the relative velocity of Q with respect to P has components of magnitude $\rho_1 \tilde{V}^1_{,1}\, dr^1$; $\rho_2\, \tilde{V}^2_{;1}\, dr^1$ in $r^1 ; r^2$-directions respectively. From (2.9.6.) it can be concluded that the r^1-component is $\rho_1 c\, dr^1$ and therefore positive, negative or undetermined depending on the interval corresponding to the direction of r^1 and the conditions

(2.9.3.) imposed on c by that interval.

When r^1 is oriented in interval (1), c is positive and the line PQ elongates. This means in the hodograph (see fig. 17) that the location of Q' is limited to a half space. This half space is bounded by the line p through P' perpendicular to the chord of PQ. The region for Q' (shaded in fig. 17) is on the side of p, corresponding to the side occupied by the endpoint Q on the line PQ. When r^1 is oriented, such as to lie in interval (3), c is negative and the location of Q' is limited to the half space bound by the same line p, but on the other side. These considerations indicate how to construct the range of velocities possible in a point of the x;y -plane by limiting the region in the hodograph, where its image is allowed to lie. Because of the conditions (2.9.3.) only lines r^1 in the intervals (1) and (3) can contribute valuable information in this respect.

However, although it is necessary, that locations of the image in the hodograph is such, that c is positive or negative depending on the interval, this is not yet sufficient. This follows from (2.9.3.) by considering f.i. the case of interval (1), where $\alpha > 0$ and , $\beta > 0$. Introduced in (2.9.2.) the requirement (2.1.1.) shows, that it is necessary for c to be positive. However, a positive value of c means that

$$\alpha a + \beta b \geqslant 0 \tag{2.9.7.}$$

and this implies

$$a \geqslant -\beta b/\alpha .$$

Since all three quantities α ; β ; b are positive in this case, it follows that there exist negative values of a that satisfy $c \geqslant 0$. By writing (2.9.7.) as $b \geqslant -\alpha a/\beta$, the possibility is shown that also b can be negative. Therefore satisfying $c \geqslant 0$ is no guarantee, that a and b are both positive separately.

In order to guarantee, that both a and b are positive, it is necessary to execute two analyses in the interval (1) for j, indicated by t^1 and t^2, here.

Analysis t^1: Take $\alpha > 0$; $\beta = 0$. This is obtained by having $j_{t^1} = -\pi/4 + \phi/2 + k\pi$. Since now c equals αa, it is enough to require $c \geqslant 0$ in order to guarantee, that a is positive.

Analysis t^2 : Take $\alpha = 0$; $\beta > 0$. This is obtained by having $j_{t^2} = +\pi/4 - \phi/2 + k\pi$. Since now c equals βb, it is enough to require $c \geqslant 0$ in order to guarantee, that b is positive.

The same can be done in interval (3), where c must be negative. The two required analyses are called s^1; s^2.

Analysis s^1: Take $\alpha = 0$; $\beta < 0$.
Then $j_{s^1} = -\pi/4 - \phi/2 + k\pi$.

Analysis s^2: Take $\alpha < 0$; $\beta = 0$.
Then $j_{s^2} = +\pi/4 + \phi/2 + k\pi$.

By requiring $c \leqslant 0$ in both analyses s^1 ; s^2, respectively, it is guaranteed that b is positive and a is positive, respectively.

Graphically the four analyses t^1; t^2; s^1; s^2 result in drawing four boundary lines in the hodograph. Let us consider the problem to determine the velocity in a point C of the x ; y -plane, when the velocities are known in four points A ; D ; E ; B which are connected to C by a s^1 ; t^1 ; t^2 ; s^2 line, respectively (see fig. 18). These four velocities are represented in the hodograph by the four image points A′ ; D′ ; E′ ; B′ .

The analysis consists in drawing lines a ; d ; e ; b in the hodograph plane, perpendicular respectively to the chords of the lines AC ; DC ; EC ; BC in the x ; y-plane. In order to satisfy the conditions (2.9.3.), the possible location for C′ in the hodograph is the shaded area in figure 18 enclosed by the lines a;d;e;b.

Instead of one unique solution for the velocity in C, the analysis results in a possible region for the image point C′, which stands for a possible range of velocities in C.

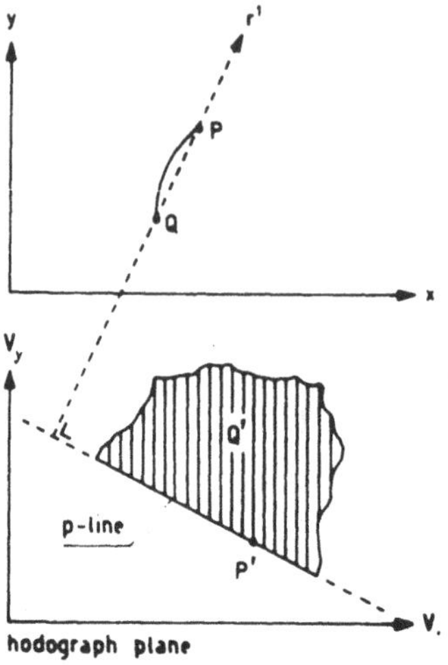

Fig. 17 – Point Q has velocity, represented by a point Q' in the hodograph plane, above the boundary line p when r^1 is in interval (1).

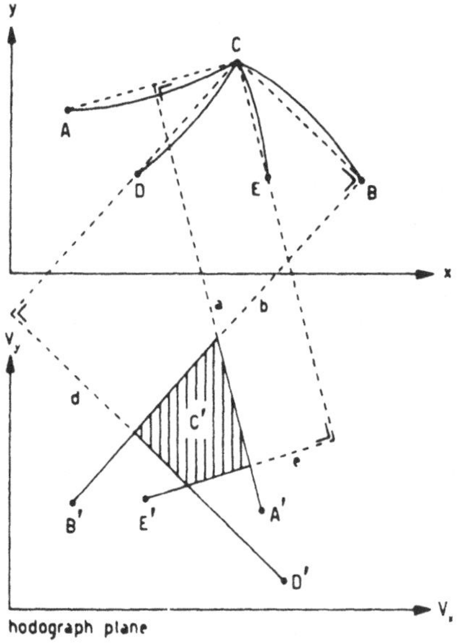

Fig. 18 – Velocity in C is represented by a point C', whose location in the hodograph plane is limited to a region enclosed by the four lines d, a, b, e.

REFERENCES

[1] Drescher, A. "An Experimental Investigation of Flow Rules for Granular Materials using Optically Sensitive Glass Particles", Geotechnique, Vol. 26, no. 4, pp. 591-601.

[2] De Josselin de Jong, G. (1958), "The Undefiniteness in Kinematics for Friction Materials", Proc. Conf. Earth Pressure Problems, Brussels, Vol. 1, pp. 55-70.

[3] De Josselin de Jong, G. (1959), "Statics and Kinematics in the Failable Zone of a Granular Material", Thesis, University, Delft, the Netherlands.

[4] De Josselin de Jong, G. (1971), "The Double Sliding, Free Rotating Model for Granular Assemblies", Geotechnique, Vol. 21, no. 2, pp. 155-163.

[5] De Josselin de Jong, G. (1977, a) "Mathematical Elaboration of the Double Sliding, Free Rotating Model". Archiwum Mechaniki Stosowanej, Vol. 29, no. 4, pp. 561-591, Warszawa.

[6] De Josselin de Jong, G. (1977, b) "Constitutive Relations for the Flow of a Granular Assembly in the Limit State of Stress". Speciality Session 9, pp. 87-95, Proc. Conf. Soil Mech. & Found. Eng., Tokyo.

[7] Sokolovskii, V.V. (1965), "Statics of Granular Media", book translated from the second Russian edition 1960, Pergamon Press.

[8] Vermeer, P.A., (1973), Private communication.

NON-ASSOCIATED FLOW RULES IN DESCRIPTION OF PLASTIC FLOW OF GRANULAR MATERIALS

Z. Mróz and Cz. Szymański
Institute of Fundamental Technological Research, Warsaw, Poland

Introduction

In most engineering problems related to determination of limit capacity of soil foundation and its interaction with a structure, only static field was considered. Starting from the Coulomb yield condition, the stress field was determined in a more or less accurate way and the estimate of limit safety factor was found. However, the knowledge of the associated velocity field is indispensable in many cases when the problems are statically indeterminate and the solution depends on velocity boundary conditions. A separate class of problems is constituted by such cases as flow of granular materials in silos and chemical reactors, during transport or in earth-working machines (cutting, filling, bulldozing, etc.) In these problems, the mechanism of flow is of primary importance and the constitutive law used in the analysis may essentially affect the solution.

In this chapter, we neglect the effects of hardening or softening of the material and assume that the limit or plastic state is described by the yield condition which does not vary with deformation. For the Coulomb yield condition, we have

$$f(\sigma) = \frac{1}{2}(\sigma_1 - \sigma_3) + \frac{1}{2}(\sigma_1 + \sigma_3)\sin\varphi - c\cos\varphi = 0. \tag{1}$$

where $\sigma_1 > \sigma_2 > \sigma_3$ are principal stresses and c, φ denote cohesion and the angle of internal friction. The alternative yield condition has the form (Mises-Schleicher yield condition)

$$f(\sigma) = J_2'^{\,1/2} + \alpha J_1 - k = 0, \tag{2}$$

where $J_2' = 1/2\, s_{ij}\, s_{ij}$, $J_1 = \sigma_{kk}$, $s_{ij} = \sigma_{ij} - 1/3\, \sigma_{kk}\, \delta_{ij}$ and φ, k are material constants.

Let us associate the flow rule with these yield conditions assuming the material behaviour as perfectly plastic. Let the total strain rate be decomposed into elastic and plastic parts

$$\dot{\underset{\sim}{\epsilon}} = \dot{\underset{\sim}{\epsilon}}^{e} + \dot{\underset{\sim}{\epsilon}}^{p}.$$

The simplest assumption would be to associate $\dot{\underset{\sim}{\epsilon}}^p$ with the yield condition $f = 0$ by the plastic potential flow rule, that is

$$(4) \qquad \dot{\underset{\sim}{\epsilon}}^p = \dot{\lambda}\,\mathrm{grad}\, f(\underset{\sim}{\sigma}) = \dot{\lambda}_1\, \underset{\sim}{n}_f$$

where $\dot{\lambda} > 0$ is the proportionality factor and $\dot{\lambda}_1 = \lambda\, |\mathrm{grad}\, f(\underset{\sim}{\sigma})|$; $\underset{\sim}{n}_f$ is the unit vector normal to the yield surface $f(\underset{\sim}{\sigma}) = 0$. A more general flow rule is obtained by assuming that the plastic potential does not coincide with the yield condition, thus

$$(5) \qquad \dot{\underset{\sim}{\epsilon}}^p = \dot{\lambda}\,\mathrm{grad}\, g(\underset{\sim}{\sigma}) = \dot{\lambda}_1\, \underset{\sim}{n}_g$$

where $g(\underset{\sim}{\sigma})$ is different from $f(\underset{\sim}{\sigma})$. Similarly as in (4), now $\underset{\sim}{n}_g = \mathrm{grad}\, g / |\mathrm{grad}\, g|$. In particular, when $g(\underset{\sim}{\sigma}) = J_2'$, from (5), we obtain

$$(6) \qquad \dot{\underset{\sim}{\epsilon}}^p = \dot{\lambda}\, \underset{\sim}{s}\ , \quad \mathrm{tr}\, \dot{\underset{\sim}{\epsilon}}^p = 0\ ,$$

and the plastic flow is incompressible. The associated flow rule (4) was proposed by Drucker and Prager (1), whereas more general non-associated flow law (5) was derived by Melan (2). The incompressibility assumption following from (6) was proposed by Hill (3), Ishlinski (4), Jenike and Shield (5) and others. Let us note that both (4) and (5) predict coincidence of principal axes of stress and strain rate tensors. In fact, if $f(\underset{\sim}{\sigma})$ and $g(\underset{\sim}{\sigma})$ are isotropic functions of stress, the flow rules (4) and (5) predict coincidence of axes of $\underset{\sim}{\sigma}$ and $\dot{\underset{\sim}{\epsilon}}^p$. In what follows, we shall call (4) and (5) *coaxial* or isotropic flow rules.

A more general situation occurs when besides stress state, the plastic strain or stress rate affect the flow of the granular material. The flow rules can now be expressed as follows

$$(7) \qquad \dot{\underset{\sim}{\epsilon}}^p = \dot{\lambda}\, \underset{\sim}{f}_1(\underset{\sim}{\sigma}, \underset{\sim}{\epsilon}^p)\ ,$$

and

$$\dot{\underset{\sim}{\epsilon}}^{p} = \dot{\underset{\sim}{\lambda}}\, f_2(\underset{\sim}{\sigma}, \overset{\nabla}{\underset{\sim}{\sigma}}), \tag{8}$$

where $\overset{\nabla}{\underset{\sim}{\sigma}}$ denotes the Zaremba-Jaumann corotational stress rate referred to axes rotating with the element. Now, the tensors $\dot{\epsilon}^{p}$ and $\underset{\sim}{\sigma}$ need no longer be coaxial although $\underset{\sim}{f}_1$ and $\underset{\sim}{f}_2$ should be isotropic functions of their arguments.

In the next section, we shall discuss various flow rules for the case of plane flow of a rigid-plastic granular material. Our discussion will be based on the representation of the flow rule in the form

$$\dot{\underset{\sim}{\epsilon}}^{p} = \dot{\lambda}_1 \operatorname{grad} g_1(\underset{\sim}{\sigma}) + \dot{\lambda}_2 \operatorname{grad} g_2(\underset{\sim}{\sigma}) + \dot{\lambda}_3 \operatorname{grad} g_3(\underset{\sim}{\sigma}) \tag{9}$$

where $\dot{\lambda}_1, \dot{\lambda}_2, \dot{\lambda}_3$ are positive parameters to be specified from the additional physical assumptions. Since the plastic strain rate should be determined to within one scalar factor, two additional equations should be added to (9) in order to make plastic flow uniquely determined. In particular, we obtain some flow rules already formulated, but new possibilities will be envisaged and through uniform discussion, better understanding of validity of different hypotheses will be illustrated. In Section 4, the application of associated and non-associated flow rules will be discussed.

2. Discussion of Flow Rules for Plane Deformation

2.1 General Representation

Consider a perfectly plastic body for which the Coulomb yield criterion (1) applies. Our discussion in this Section will be limited to plane plastic flow, when

$$v_x = u(x, y, t)\,, \quad v_y = v(x, y, t)\,, \quad v_z = 0 \tag{10}$$

and the plane of flow is the principal plane in which σ_1 and σ_3 are major and minor principal stresses; the stress σ_2 normal to the plane of flow is the intermediate principal stress. Thus

$$\begin{aligned} &\sigma_x = \sigma_x(x, y, t)\,, \quad \sigma_y = \sigma_y(x, y, t)\,, \quad \tau_{xy} = \tau_{xy}(x, y, t)\,, \\ &\sigma_z = \sigma_z(x, y, t)\,, \quad \tau_{zx} = \tau_{zy} = 0\,, \quad \sigma_1 > \sigma_2 > \sigma_3\,. \end{aligned} \tag{11}$$

The Coulomb yield condition is now expressed as follows

$$f(\sigma) = [\tfrac{1}{4}(\sigma_x - \sigma_y)^2 + \tau_{xy}^2]^{1/2} + \tfrac{1}{2}(\sigma_x + \sigma_y)\sin\varphi - c\cos\varphi = 0\,, \tag{12}$$

and we assume the tensile stresses as positive, Fig. 1. For further analysis, it is convenient to introduce new stress variables

$$p = \tfrac{1}{2}(\sigma_x + \sigma_y)\,, \quad q = \tfrac{1}{2}(\sigma_x - \sigma_y)\,, \quad r = \tau_{xy}\,, \tag{13}$$

and the associated strain rates

$$\begin{aligned} &\dot{\epsilon}_p = \dot{\epsilon}_x^p + \dot{\epsilon}_y^p = \frac{\partial u}{\partial x} + \frac{\partial v}{\partial y}\,, \quad \dot{\epsilon}_q = (\dot{\epsilon}_x^p - \dot{\epsilon}_y^p) = \frac{\partial u}{\partial x} - \frac{\partial v}{\partial y}\,, \\ &\dot{\epsilon}_r = \dot{\gamma}_{xy}^p = \frac{\partial u}{\partial y} + \frac{\partial v}{\partial x}\,, \end{aligned} \tag{14}$$

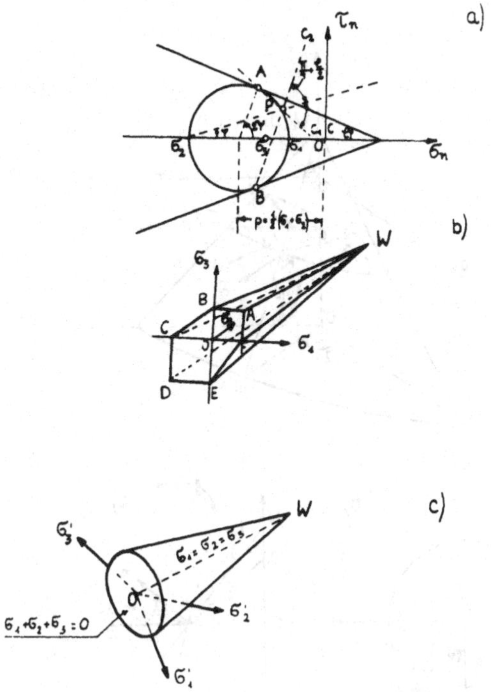

Fig. 1

so that the specific dissipation rate is expressed as follows

$$D = \sigma_x \dot{\epsilon}^p_x + \sigma_y \dot{\epsilon}^p_y + \tau_{xy} \dot{\gamma}^p_{xy} = p \dot{\epsilon}_p + q \dot{\epsilon}_q + r \dot{\epsilon}_r \,. \tag{15}$$

The yield condition now takes the form

$$f(q, r, p) = (q^2 + r^2)^{1/2} + p \sin\varphi - c \cos\varphi = 0 \,. \tag{16}$$

In the space (q, r, p), the yield condition (16) is represented by a circular cone with the axis Op and the vertex W(O, O, c cot φ), Fig. 2a. The intersection of the cone by the plane p = const. is shown in Fig. 2b and Fig. 2c shows the intersection by the meridional plane. We have

$$(q^2 + r^2)^{1/2} = R(p) \,, \quad R(p) = c \cos\varphi - p \sin\varphi \tag{17}$$

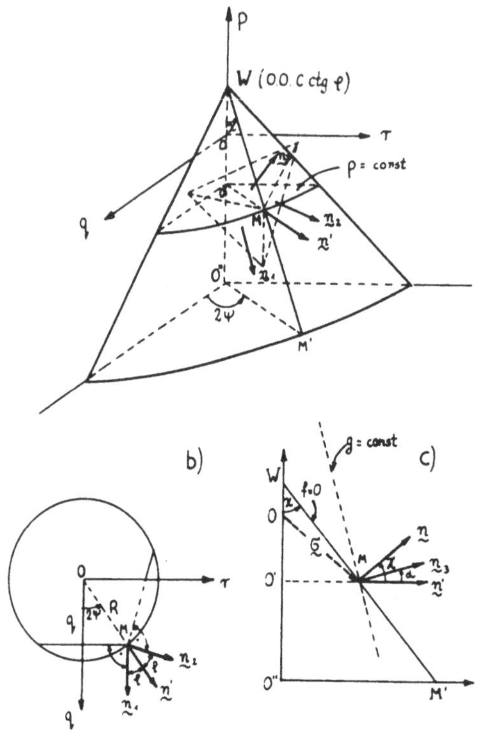

Fig. 2

and $\operatorname{tg}\chi = \sin\varphi$, $0 \leqslant \chi \leqslant \pi/4$.

In order to discuss sufficiently general constitutive law, assume that the plastic potential is a singular vertex obtained by intersection of three planes with normal unit vectors $\underset{\sim}{n}_1$ $\underset{\sim}{n}_2$ $\underset{\sim}{n}_3$, thus

$$(18) \qquad \dot{\underset{\sim}{\epsilon}}^p = \dot{\lambda}_1 \underset{\sim}{n}_1 + \dot{\lambda}_2 \underset{\sim}{n}_2 + \dot{\lambda}_3 \underset{\sim}{n}_3 ,$$

where $\dot{\lambda}_1$, $\dot{\lambda}_2$, $\dot{\lambda}_3$ are three positive functions whose meaning will be discussed in he following. Assume that the vectors $\underset{\sim}{n}_1$ and $\underset{\sim}{n}_2$ make equal angles φ with the meridional plane and the vector $\underset{\sim}{n}_3$ is inclined at the angle $\chi - \alpha$ to the exterior cone normal. Thus, in the (q, p, r)-system, we have

$$
\begin{aligned}
\underset{\sim}{n}_1 &= \cos(2\psi - \varphi)\,,\ \sin(2\psi - \varphi)\,,\ 0\ , \\
\underset{\sim}{n}_2 &= \cos(2\psi + \varphi)\,,\ \sin(2\psi + \varphi)\,,\ 0\,, \qquad (19)\\
\underset{\sim}{n}_3 &= \cos\alpha \cos 2\psi\,,\ \cos\alpha \sin 2\psi\,,\ \sin\alpha\ ,
\end{aligned}
$$

where

$$
\mathrm{tg}\, 2\psi = \frac{2\tau_{xy}}{\sigma_x - \sigma_y} = \frac{r}{q}\ , \qquad (20)
$$

and the flow rule gives

$$
\begin{aligned}
\dot{\epsilon}_q &= \dot{\lambda}_1 \cos(2\psi - \varphi) + \dot{\lambda}_2 \cos(2\psi + \varphi) + \dot{\lambda}_3 \cos\alpha \cos 2\psi\ , \\
\dot{\epsilon}_r &= \dot{\lambda}_1 \sin(2\psi - \varphi) + \dot{\lambda}_2 \sin(2\psi + \varphi) + \dot{\lambda}_3 \cos\alpha \sin 2\psi\ , \qquad (21)\\
\dot{\epsilon} &= \dot{\lambda}_3 \sin\alpha
\end{aligned}
$$

with $\dot{\lambda}_1 > 0$, $\dot{\lambda}_2 > 0$, $\dot{\lambda}_3 > 0$. Moreover, [illegible] vanish if $f = 0$, $\dot{f} < 0$ or $f < 0$. The flow rule (18) with a hidden corner on a smooth yield surface was discussed in [14].

Since $q = R \cos 2\psi$, $r = R \sin 2\psi$, the flow rule (21) can also be expressed as follows

$$
\dot{\epsilon}_q = \dot{a} \cos 2\psi + \dot{b} \sin 2\psi\,,\quad \dot{\epsilon}_r = \dot{a} \sin 2\psi - \dot{b} \cos 2\psi\,,\quad \dot{\epsilon}_p = \dot{\lambda}_3 \sin\alpha \qquad (22)
$$

or

$$\dot{\epsilon}_q = \dot{a}\,\frac{q}{R} + \dot{b}\,\frac{r}{R}\,, \qquad \dot{\epsilon}_r = \dot{a}\,\frac{r}{R} - \dot{b}\,\frac{q}{R}\,, \qquad \dot{\epsilon}_p = \dot{\lambda}_3 \sin\alpha \tag{23}$$

where

$$\dot{a} = (\dot{\lambda}_1 + \dot{\lambda}_2)\cos\varphi + \dot{\lambda}_3\cos\alpha\,, \qquad \dot{b} = (\dot{\lambda}_1 - \dot{\lambda}_2)\sin\varphi\,. \tag{24}$$

It is seen that two additional equations must be added to (21) or (23) in order to define plastic flow uniquely. First, however, let us discuss all flow rules already proposed in the literature. They can easily be derived either from (21) or from (23).

2.2. Coaxial Flow Rules

Setting $\dot{\lambda}_1 = \dot{\lambda}_2$ or $\dot{b} = 0$ in (21) or (22), we obtain

$$\dot{\epsilon}_q = \dot{a}\cos 2\psi\,, \qquad \dot{\epsilon}_r = \dot{a}\sin 2\psi\,, \quad \dot{\epsilon}_p = \dot{\lambda}_3 \sin\alpha\,, \tag{25}$$

or

$$\dot{\epsilon}_q = \dot{a}\,\frac{q}{R}\,, \qquad \dot{\epsilon}_r = \dot{a}\,\frac{r}{R}\,, \qquad \dot{\epsilon}_p = \dot{\lambda}_3 \sin\alpha \tag{26}$$

and

$$\frac{\dot{\epsilon}_q}{\dot{\epsilon}_r} = \frac{r}{q} \tag{27}$$

The condition (27) implies coincidence of axes of stress and strain rate. However, since there are two parameters in (25) or (26), an additional hypothesis may be introduced that would relate $\dot{a}$ and $\dot{\lambda}_3$. Several possibilities are now considered.

(i) Setting $\dot{\lambda}_1 = \dot{\lambda}_2 = 0$, we have $\dot{a} = \dot{\lambda}_3\cos\alpha$, $\dot{b} = 0$. Denoting $\dot{\lambda} = \dot{a} = \dot{\lambda}_3\cos\alpha$, $\sin\varphi' = \mathrm{tg}\,\alpha$, we have

$$\dot{\epsilon}_q = \dot{\lambda}\cos 2\psi\,, \qquad \dot{\epsilon}_r = \dot{\lambda}\sin 2\psi\,, \qquad \dot{\epsilon}_p = \dot{\lambda}\,\mathrm{tg}\,\alpha = \dot{\lambda}\sin\varphi'\,, \tag{28}$$

or

$$\dot{\epsilon}_q = \dot{\lambda}\frac{q}{R}, \qquad \dot{\epsilon}_r = \dot{\lambda}\frac{r}{R}, \qquad \dot{\epsilon}_p = \dot{\lambda}\sin\varphi' . \tag{29}$$

The flow rule (29) can be obtained from (5) by assuming the plastic potential in the form

$$g(p, q, r) = (q^2 + r^2)^{1/2} + p\sin\varphi' - c\cos\varphi' . \tag{30}$$

Now (29) is equivalent to the potential relations

$$\dot{\epsilon}_q = \dot{\lambda}\frac{\partial g}{\partial q}, \qquad \dot{\epsilon}_r = \dot{\lambda}\frac{\partial g}{\partial r}, \qquad \dot{\epsilon}_p = \dot{\lambda}\sin\varphi' = \dot{\lambda}\frac{\partial g}{\partial p} . \tag{31}$$

In particular, when $\dot{\lambda}_1 = \dot{\lambda}_2 = 0$, $\alpha = \chi$, (31) passes to the associated flow rule

$$\dot{\epsilon}_q = \dot{\lambda}\frac{q}{R} = \dot{\lambda}\frac{\partial f}{\partial q}, \quad \dot{\epsilon}_r = \dot{\lambda}\frac{r}{R} = \dot{\lambda}\frac{\partial f}{\partial r}, \quad \dot{\epsilon}_p = \dot{\lambda}\sin\varphi = \dot{\lambda}\frac{\partial f}{\partial p} \tag{32}$$

which was applied by R.T. Shield in studying plane plastic flow (6).
(ii) Setting $\dot{a} = \dot{\lambda}$, $\dot{b} = 0$, $\dot{\lambda}_3 \sin\alpha = 0$, we have

$$\dot{\epsilon}_q = \dot{\lambda}\cos 2\psi, \qquad \dot{\epsilon}_r = \dot{\lambda}\sin 2\psi, \qquad \dot{\epsilon}_p = 0, \tag{33}$$

and the plastic flow is incompressible. Thus (33) is equivalent to (6) which was postulated by numerous authors (3-5).

2.3. *Non-coaxial Flow Rules*

Non-coaxiality of stress and strain-rate tensors follows from the assumption that the flow mechanism consists of glides occuring along one or the other stress characteristics or as a combined glide on two characteristics. Considering the vertex representation shown in Fig. 2, it may be conceived that for incompressible flow the two planes with normals $\underset{\sim}{n}_1$ and $\underset{\sim}{n}_2$ represent two gliding mechanisms on which purely dilatational flow is superposed governed by the potential plane with the normal $\underset{\sim}{n}_3$.

The angle between principal directions $\underset{\sim}{\sigma}_1$ and $\dot{\underset{\sim}{\epsilon}}_1$ is expressed as follows, see Fig. 3

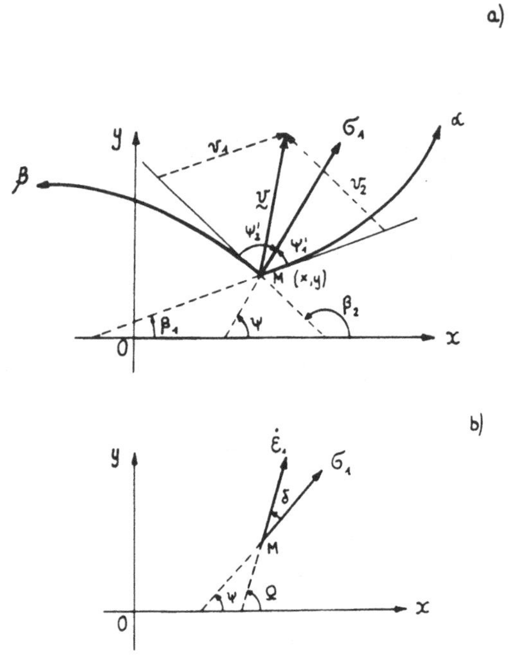

Fig. 3

(34) $$\delta = \Omega - \psi ,$$

where

(35) $$\operatorname{tg} 2\psi = \frac{2\tau_{xy}}{\sigma_x - \sigma_y} = \frac{r}{q} , \qquad \operatorname{tg} 2\Omega = \frac{\dot{\gamma}_{xy}}{\dot{\epsilon}_x - \dot{\epsilon}_y} .$$

Hence

$$\operatorname{tg}2\delta = \frac{\operatorname{tg}2\Omega - \operatorname{tg}2\psi}{1 + \operatorname{tg}2\Omega\,\operatorname{tg}2\psi} = \frac{q\dot{\epsilon}_r - r\dot{\epsilon}_q}{q\dot{\epsilon}_q - r\dot{\epsilon}_r} . \tag{36}$$

Using (21) or (22), Eq. (36) can be presented in the form

$$\operatorname{tg}2\delta = -\frac{\dot{b}}{\dot{a}} = \frac{(\dot{\lambda}_2 - \dot{\lambda}_1)\sin\varphi}{(\dot{\lambda}_2 + \dot{\lambda}_1)\cos\varphi + \dot{\lambda}_3\cos\alpha} . \tag{37}$$

Let us note that for $\dot{\lambda}_1 \geqslant 0$, $\dot{\lambda}_2 \geqslant 0$ and $\dot{\lambda}_3 \geqslant 0$, we have

$$-\operatorname{tg}\varphi \leqslant \operatorname{tg}2\delta \leqslant \operatorname{tg}\varphi \tag{38}$$

and

$$-\frac{1}{2}\varphi \leqslant \delta \leqslant +\frac{1}{2}\varphi , \tag{39}$$

thus the equality sign occurs for $\dot{\lambda}_3 = 0$, $\dot{\lambda}_2 = 0$ and $\dot{\lambda}_1 > 0$ or $\dot{\lambda}_3 = 0$, $\dot{\lambda}_1 = 0$ and $\dot{\lambda}_2 > 0$. Now, let us express two known hypotheses proposed by Geniev (11), Takagi (7) and de Josselin de Jong (8).

(i) Geniev model: it is assumed that plastic flow occurs with extremal values of δ that is $\delta = -1/2\,\varphi$ or $\delta = +1/2\,\varphi$. This assumption can be expressed as follows

$$\dot{\lambda}_2 = \dot{\lambda}_3 = 0,\ \dot{\lambda}_1 > 0 \quad \rightarrow \quad \delta = -\frac{1}{2}\varphi,\ \dot{\epsilon}_p = 0, \tag{40}$$

or

$$\dot{\lambda}_1 = \dot{\lambda}_3 = 0,\ \dot{\lambda}_2 > 0 \quad \rightarrow \quad \delta = +\frac{1}{2}\varphi,\ \dot{\epsilon}_p = 0. \tag{41}$$

and the flow rules take the form

$$(42)\qquad \dot{\epsilon}_q = \dot{\lambda}_1 \cos(2\psi - \varphi)\,, \quad \dot{\epsilon}_r = \dot{\lambda}_1 \sin(2\psi - \varphi)\,, \quad \dot{\epsilon}_p = 0\,, \quad \delta = -\frac{1}{2}\varphi\,,$$

$$(43)\qquad \dot{\epsilon}_q = \dot{\lambda}_2 \cos(2\psi + \varphi)\,, \quad \epsilon_r = \dot{\lambda}_2 \sin(2\psi + \varphi)\,, \quad \dot{\epsilon}_p = 0, \quad \delta = +\frac{1}{2}\varphi\,.$$

It is not evident, however, which set of relations should be used and in general two flow mechanisms are possible. In the next section, we shall illustrate applicability of (42) and (43) in some particular cases.

(ii) de Josselin de Jong model. Assume the combined shear along two stress characteristics as the mechanism of incompressible flow. Let $\dot{\lambda}_3 = 0$, then from (37) it follows that

$$(44)\qquad \mathrm{tg}\,2\delta = \frac{\dot{\lambda}_2 - \dot{\lambda}_1}{\dot{\lambda}_2 + \dot{\lambda}_1}\,\mathrm{tg}\,\varphi = -\frac{\dot{b}}{\dot{a}}\,,$$

and for $\dot{\lambda}_1 > 0$ and $\dot{\lambda}_2 > 0$ the inequality (39) is satisfied. Since no additional assumption is introduced at this moment, the angle δ can take any value between $-1/2\,\varphi$ and $+1/2\,\varphi$. It is seen that the corresponding flow rule is incomplete since the relation between $\dot{\lambda}_1$ and $\dot{\lambda}_2$ (or $\dot{a}$ and $\dot{b}$) is lacking.

(iii) Anisotropy hypothesis. In order to find an additional relation between $\dot{\lambda}_1$ and $\dot{\lambda}_2$, it may be postulated that the state of strain affects the ratio $\dot{\lambda}_1 / \dot{\lambda}_2$. Similarly as for the strain rates, let us denote $\epsilon_q = \epsilon^p_x - \epsilon^p_y$, $\epsilon_r = \gamma^p_{xy}$ whereas $\epsilon_p = 0$ since plastic deformation is assumed to be incompressible. Let us consider the representation

$$(45)\qquad \underset{\sim}{\epsilon} = \Lambda_1 \underset{\sim}{n}_1 + \Lambda_2 \underset{\sim}{n}_2\,,$$

or

$$(46)\qquad \begin{aligned} \epsilon_q &= \Lambda_1 \cos(2\psi - \varphi) + \Lambda_2 \cos(2\psi + \varphi)\,, \\ \epsilon_r &= \Lambda_1 \sin(2\psi - \varphi) + \Lambda_2 \sin(2\psi + \varphi)\,. \end{aligned}$$

Let us note that now Λ_1 and Λ_2 may be both negative and positive since the strain vector on the plane (q,r) may have an arbitrary orientation. Now, the hypothesis is made that

$$\frac{\dot{\lambda}_1}{\dot{\lambda}_2} = \frac{|\Lambda_1|}{|\Lambda_2|} \tag{47}$$

that is ratio of rates of shears along the two characteristic directions is proportional to strain components along the directions of characteristics. From (46), we find

$$|\Lambda_1| = \frac{|\epsilon_q \sin(2\psi + \varphi) - \epsilon_r \cos(2\psi + \varphi)|}{\sin 2\varphi}$$
$$|\Lambda_2| = \frac{|\epsilon_r \cos(2\psi - \varphi) - \epsilon_q \sin(2\psi - \varphi)|}{\sin 2\varphi} \tag{48}$$

and the flow rule takes the form

$$\dot{\underline{\epsilon}}^p = \dot{\lambda}_1 [\underline{n}_1 + A\underline{n}_2] , \tag{49}$$

where

$$A = \frac{|\Lambda_2|}{|\Lambda_1|} = \frac{|\sin(2\theta - 2\psi + \varphi)|}{|\sin(2\psi + \varphi - 2\theta)|} , \quad \mathrm{tg}\, 2\theta = \frac{\epsilon_r}{\epsilon_q} , \quad 0 < A < \infty . \tag{50}$$

Let us note that for $2\theta = 2\psi - \varphi$, we have $A = 0$ whereas for $2\theta = 2\psi + \varphi$, there is $A = \infty$. In these two limiting cases we arrive at the Geniev hypothesis. It follows from (50) that only relative orientation of principal axes of strain with respect to principal stress axes affects the flow rule. Thus the plastic flow may be studied by determining velocity field and recording orientation of the principal strain axes for small increments in deformation. Note that for $\varphi = 0$, Eq. (50) yields $A = 1$ and (49) is equivalent to the coaxial flow rule, usually applied in studying plastic flow of metals. Both linear and non-linear strain measures can be assumed in (49).

2.4. Velocity Characteristics

The strain equations composed of the two equilibrium equations and the yield conditions (1) or (2) are well investigated and need not be discussed here, see Sokolovski (12). The stress characteristics are inclined at the angles $\psi \pm (\pi/4 + \varphi/2)$ to the x-axis and the stress field is coupled with the velocity field only through the boundary conditions which involve usually both surface tractions and velocities.

Let us now discuss the velocity equations for plane flow assuming a general representation of the flow rule in the form (18) or (22). It is easy to see that from (22) the following two equalities can be formed

$$
\begin{aligned}
&(\dot{a}\sin 2\psi - \dot{b}\cos 2\psi)\dot{\epsilon}_q - (\dot{a}\cos 2\psi + \dot{b}\sin 2\psi)\dot{\epsilon}_r = 0,\\
(51)\qquad &\\
&(\dot{a}\cos 2\psi + \dot{b}\sin 2\psi)\dot{\epsilon}_p - \dot{\lambda}_3 \sin\alpha\, \dot{\epsilon}_q = 0\,,
\end{aligned}
$$

and substituting (14), we find

$$
\begin{aligned}
&(\sin 2\psi - \frac{\dot{b}}{\dot{a}}\cos 2\psi)(\frac{\partial u}{\partial x} - \frac{\partial v}{\partial y}) - (\cos 2\psi + \frac{\dot{b}}{\dot{a}}\sin 2\psi)(\frac{\partial u}{\partial y} + \frac{\partial v}{\partial x}) = 0\,,\\
(52)\qquad &\\
&(\cos 2\psi + \frac{\dot{b}}{\dot{a}}\sin 2\psi)(\frac{\partial u}{\partial x} + \frac{\partial v}{\partial y}) - \frac{\dot{\lambda}_3}{\dot{a}}\sin\alpha(\frac{\partial u}{\partial y} + \frac{\partial v}{\partial x}) = 0\,.
\end{aligned}
$$

The set (52) is hyperbolic provided

$$
(53)\qquad \dot{a}^2 + \dot{b}^2 - \dot{\lambda}_3^2 \sin^2\alpha > 0\,.
$$

and possesses two families of real characteristics coinciding with the lines of zero extension rates. In fact, expressing the set (52) in the curvilinear characteristics coordinates α, β, we obtain (see Fig. 3)

$$\dot{\epsilon}_\alpha = \frac{\partial v_1}{\partial s_\alpha} + \cos 2\mu_k \frac{\partial v_2}{\partial s_\alpha} - v_2 \sin 2\mu_k \frac{\partial \beta_2}{\partial s_\alpha} = 0 ,$$

$$\dot{\epsilon}_\beta = \frac{\partial v_2}{\partial s_\beta} + \cos 2\mu_k \frac{\partial v_1}{\partial s_\beta} + v_1 \sin 2\mu_k \frac{\partial \beta_1}{\partial s_\beta} = 0 , \tag{54}$$

where $\beta_1 = \psi - \psi_1'$, $\beta_2 = \psi - \psi_2'$ and $2\mu_k = \beta_2 - \beta_1 = \psi_1' - \psi_2'$, where

$$\cos 2\mu_k = - \frac{\dot{\lambda}_3 \sin\alpha}{(\dot{a}^2 + \dot{b}^2)^{1/2}} , \quad \mathrm{tg}\,\psi_1' = \frac{\dot{b} + (\dot{a}^2 + \dot{b}^2 - \dot{\lambda}_3^2 \sin^2\alpha)^{1/2}}{\dot{a} - \dot{\lambda}_3 \sin\alpha} ,$$

$$\mathrm{tg}\,\psi_2' = \frac{\dot{b} - (\dot{a}^2 + \dot{b}^2 - \dot{\lambda}_3^2 \sin^2\alpha)^{1/2}}{\dot{a} - \dot{\lambda}_3 \sin\alpha} , \tag{55}$$

and β_1, β_2 denote the angles of characteristic directions at the point M with the x-axis; V_1 and V_2 are velocity components along the lines α and β and ψ_1', ψ_2' are the angles between the lines α and β and the trajectory of principal stress σ_1, Fig. 3.

For any flow rule the characteristic relations can easily be derived from the set (52) or (54). Consider first incompressible flow $\dot{\epsilon}_p = \dot{\lambda}_3 = 0$. The set (52) now takes the form

$$(\sin 2\psi - \frac{\dot{b}}{\dot{a}} \cos 2\psi) (\frac{\partial u}{\partial x} - \frac{\partial v}{\partial y}) - (\cos 2\psi + \frac{\dot{b}}{\dot{a}} \sin 2\psi) (\frac{\partial u}{\partial y} + \frac{\partial v}{\partial x}) = 0$$

$$\frac{\partial u}{\partial x} + \frac{\partial v}{\partial y} = 0 . \tag{56}$$

This set is hyperbolic and its characteristics are inclined at the angle $+ \psi_1'$ and $+ \psi_2'$ to the stress trajectory of σ_1. Now, we have

(57) $$\mathrm{tg}\psi_1' = \frac{\dot b + \sqrt{\dot a^2 + \dot b^2}}{\dot a}\,, \qquad \mathrm{tg}\psi_2' = \frac{\dot b - \sqrt{\dot a^2 + \dot b^2}}{\dot a}$$

and $2\mu_k = \pi/2$. Since the ratio $\dot b/\dot a$ is unspecified by the flow rule, the field of characteristics is not uniquely determined and the orthogonal net may take different orientation with respect to the principal stress trajectories or stress characteristics. The characteristic relations (54) take the form

(58) $$\dot\epsilon_\alpha = \frac{\partial v_1}{\partial s_\alpha} - v_2 \frac{\partial \beta_1}{\partial s_\alpha} = 0\,, \qquad \dot\epsilon_\beta = \frac{\partial v_2}{\partial s_\beta} - v_2 \frac{\partial \beta_1}{\partial s_\beta} = 0\,,$$

and $\beta_2 = \beta_1 + \pi/2$.

The case of the associated flow rule is easily rederived from (52) by setting $\dot a = \dot\lambda$, $\dot b = 0$, $\dot\lambda_3 \sin\alpha = \dot\lambda\,\mathrm{tg}\alpha = \dot\lambda \sin\varphi$ and

(59) $$\mathrm{tg}\psi_1' = \frac{\cos\varphi}{1-\sin\varphi} = \mathrm{tg}(\frac{\pi}{4} + \frac{\varphi}{2})\,, \quad \mathrm{tg}\psi_2' = -\frac{\cos\varphi}{1-\sin\varphi} = -\mathrm{tg}(\frac{\varphi}{2} + \frac{\pi}{4})$$

hence $\psi_1' = \pi/4 + \varphi/2$, $\psi_2' = -(\pi/4 + \varphi/2)$, $2\mu_k = \pi/2 + \varphi$, and the velocity and the stress characteristic coincide. The characteristic relations (54) now become

(60) $$\begin{aligned} &\frac{\partial v_1}{\partial s_\alpha} - \sin\varphi \frac{\partial v_2}{\partial s_\alpha} - v_2 \cos\varphi \frac{\partial \psi}{\partial s_\alpha} = 0\,, \\ &\frac{\partial v_2}{\partial s_\beta} - \sin\varphi \frac{\partial v_1}{\partial s_\beta} + v_2 \cos\varphi \frac{\partial \psi}{\partial s_\beta} = 0 \end{aligned}$$

and are identical to those derived by Shield (6).

The characteristic equations for Geniev theory can easily be derived from (54) and (55) by setting $\dot\lambda_3 = 0$, $\dot a = \dot\lambda_1 \cos\varphi$, $\dot b = \dot\lambda_1 \sin\varphi$ or $\dot\lambda_3 = 0$, $\dot a = \dot\lambda_2 \cos\varphi$, $\dot b = -\dot\lambda_2 \sin\varphi$. Now one family of the orthogonal net of velocity characteristics coincide with the stress characteristic.

2.5. Plastic Flow with Variable Dilatancy

So far, the dilatancy of the material was assumed to be governed by the flow rule (22) with unspecified $\dot{\lambda}_3$ or by the potential flow rules (4) and (5) for which the rate of dilation is proportional to the shearing strain rate. Now, let us introduce more specific relations expressing the rate of volume change as a function of current and critical densities, that is

$$\dot{\epsilon}^p_v = F(\frac{\rho}{\rho_c})\dot{e}^p \tag{61}$$

where $\dot{e} = (\dot{e}_{ij}\,\dot{e}_{ij})^{1/2}$ or $\dot{e} = \dot{\epsilon}_1 - \dot{\epsilon}_2$, $\dot{e}_{ij} = \dot{\epsilon}_{ij} - 1/3\epsilon_{kk}\,\delta_{ij}$; ρ and ρ_c denote the actual and the critical denstiy which is a characteristic function of the material $\rho_c = \rho_c(p)$. Let us assume the dilatancy law in the form

$$F(\frac{\rho}{\rho_c}) = \alpha(\frac{\rho}{\rho_c} - 1)^n , \tag{62}$$

where α is constant and n is an odd integer. Equation (62) predicts that $F > 0$ if $\rho > \rho_c$ $F < 0$ if $\rho < \rho_c$ and $F = 0$ for $\rho = \rho_c$. Alternatively, the dilatancy law can also be expressed as follows

$$F(\frac{\rho}{\rho_c}) = \alpha \sin(\frac{\rho}{\rho_c} - 1) . \tag{63}$$

The general coaxial flow rule can now be expressed as follows

$$\dot{\epsilon}^p_{ij} = \dot{\lambda}[s_{ij} + s\,F(\frac{\rho}{\rho_c})\,\delta_{ij}] \tag{64}$$

where $s = (s_{ij}\,s_{ij})^{1/2} = (2J_2')$. For the case of plane flow, we have

$$\dot{\epsilon}_q = 2\dot{\lambda}\,R\cos 2\psi, \quad \dot{\epsilon}_r = 2\dot{\lambda}R\sin 2\psi , \quad \dot{\epsilon}_p = 2\dot{\lambda}RF(\frac{\rho}{\rho_c}) \tag{65}$$

and this flow rule is equivalent to the assumption

$$\dot{a} = 2R\dot{\lambda}\,, \quad \dot{b} = 0\,, \quad \dot{\lambda}_3 \sin\alpha = 2R\dot{\lambda}\,F(\frac{\rho}{\rho_c}) = \dot{a}\,F(\frac{\rho}{\rho_c})\,, \tag{66}$$

and

$$\dot{\epsilon}_p = \dot{e}\,F(\frac{\rho}{\rho_c})\,, \qquad \dot{e} = \dot{\epsilon}_1^p - \dot{\epsilon}_2^p\,. \tag{67}$$

The set of velocity equations now takes the form

$$\begin{aligned} &\sin 2\psi\,(\frac{\partial u}{\partial x} - \frac{\partial v}{\partial y}) - \cos 2\psi\,(\frac{\partial u}{\partial y} + \frac{\partial v}{\partial x}) = 0\,,\\ &\cos 2\psi\,(\frac{\partial u}{\partial x} + \frac{\partial v}{\partial y}) - F(\frac{\rho}{\rho_c})\,(\frac{\partial u}{\partial y} + \frac{\partial v}{\partial x}) = 0\,. \end{aligned} \tag{68}$$

The set is hyperbolic provided $-1 < F < 1$. The angles ψ_1' and ψ_2' now are

$$\operatorname{tg}\psi_1' = \frac{\sqrt{1-F^2}}{1-F}\,, \qquad \operatorname{tg}\psi_2' = -\frac{\sqrt{1-F^2}}{1-F}\,. \tag{69}$$

In the particular case, when $F(\rho/\rho_c) = \sin\varphi_k$, we obtain

$$\psi_1' = \pi/4 + \varphi_k/2\,, \qquad \psi_2' = -(\frac{\pi}{4} + \varphi_k/2) \tag{70}$$

and the characteristic equations are

$$\begin{aligned} &\frac{\partial v_1}{\partial s_\alpha} - \sin\varphi_k\,\frac{\partial v_2}{\partial s_\alpha} - v_2\cos\varphi_k\,\frac{\partial \beta_2}{\partial s_\alpha} = 0\,,\\ &\frac{\partial v_2}{\partial s_\beta} - \sin\varphi_k\,\frac{\partial v_1}{\partial s_\beta} + v_1\cos\varphi_k\,\frac{\partial \beta_1}{\partial s_\beta} = 0\,. \end{aligned} \tag{71}$$

The stress and velocity characteristics coincide only when $F = \sin\varphi$, that is for the associated flow rule. For the variable dilatancy model, the velocity characteristics change their angle in the course of deformation, tending to orthogonality in the critical state, $\rho = \rho_c = F(\rho/\rho_c) = F(1) = 0$. The plastic flow may now be studied by determining velocity field and density variation at each incremental step of the deformation process and finding new characteristic directions.

3. Flow Rules Affected by the Stress Rate

In this section, we shall consider a more general flow rule for which the plastic strain rate depends also on the stress rate. This form of flow rule is not typical for perfectly plastic solids, for which the stress state alone defines the direction of the plastic strain rate. Formally, we assume here that besides the strain rate component normal to the plastic potential surface, there is another plastic strain rate component related to the stress rate. Note that for a perfectly plastic material the stress rate vector should be directed tangentially to the yield surface. It will be shown that dependence on the stress rate can be interpreted by requiring that combined shear mechanism occurs along stress characteristics which change their orientation due to stress redistribution. In particular, the velocity equations derived by Spencer (13) will be shown to be a particular case of more general relations valid for both dilating and incompressible materials.

Starting from the assumption $\dot{\underset{\sim}{\epsilon}} = \dot{\underset{\sim}{\epsilon}}^e + \dot{\underset{\sim}{\epsilon}}^p$ let us assume that the elastic strain rate component is described as follows

$$\dot{\underset{\sim}{\epsilon}}^e = \eta_1 \overset{\triangledown}{\underset{\sim}{\sigma}} - \eta_2 (\mathrm{tr}\, \overset{\triangledown}{\underset{\sim}{\sigma}}) \underset{\sim}{I}, \quad \text{for } f(\underset{\sim}{\sigma}) < 0 \ \text{ or } \ f = 0, (\overset{\triangledown}{\underset{\sim}{\sigma}} \cdot \underset{\sim}{n}) < 0 \tag{72}$$

where

$$\begin{aligned} \overset{\triangledown}{\sigma}_{ij} &= \dot{\sigma}_{ij} + \sigma_{im}\,\omega_{jm} + \sigma_{jm}\,\omega_{mi}\,, \\ \dot{\sigma}_{ij} &= \frac{\partial \sigma_{ij}}{\partial t} + v_k \frac{\partial \sigma_{ij}}{\partial x_k}\,, \quad \omega_{ij} = \frac{1}{2}(v_{i,j} - v_{j,i})\,. \end{aligned} \tag{73}$$

Here $\overset{\triangledown}{\sigma}_{ij}$ denotes the corotational Zaremba stress rate referred to coordinate axes rotating with the element, $\dot{\sigma}_{ij}$ denotes the material stress rate and ω_{ij} is the element spin. Since $\dot{\sigma}_{ij}\, n_j = \overset{\triangledown}{\sigma}_{ij}\, n_j$, the unloading criterion can be expressed either in terms of $\dot{\sigma}_{ij}$ or $\overset{\triangledown}{\sigma}_{ij}$. The constants η_1 and η_2 define elastic properties of the material. The plastic strain rate is assumed in the form

$$\dot{\underset{\sim}{\epsilon}}^p = \dot{\lambda}\, \mathrm{grad}\, g(\underset{\sim}{\sigma}) + \aleph_1 \overset{\triangledown}{\underset{\sim}{\sigma}} - \aleph_2 (\mathrm{tr}\, \overset{\triangledown}{\sigma}) \underset{\sim}{I}, \quad f(\underset{\sim}{\sigma}) = \dot{f}(\underset{\sim}{\sigma}) = 0 \tag{74}$$

where $\dot{\lambda} > 0$. Thus, the total strain rate can be expressed as follows

$$\dot{\underset{\sim}{\epsilon}} = \dot{\lambda}\,\mathrm{grad}\,g(\underset{\sim}{\sigma}) + \vartheta_1 \overset{\triangledown}{\underset{\sim}{\sigma}} - \vartheta_2 (\mathrm{tr}\,\sigma)\underset{\sim}{I} \;, \tag{75}$$

where

$$\begin{aligned} &\vartheta_1 = \eta_1 , \quad \vartheta_2 = \eta_2 , \qquad \text{for } f(\underset{\sim}{\sigma}) < 0 \text{ or } f(\underset{\sim}{\sigma}) = 0,\ \dot{f}(\underset{\sim}{\sigma}) < 0 , \\ &\vartheta_1 = \eta_1 + \aleph_1 , \quad \vartheta_2 = \eta_2 + \aleph_2 , \qquad \text{for } f(\underset{\sim}{\sigma}) = \dot{f}(\underset{\sim}{\sigma}) = 0. \end{aligned} \tag{76}$$

The presented constitutive equations are valid for any stress state. Let us now confine our discussion to the case of plane flow. Assume that

$$g(\underset{\sim}{\sigma}) = (q^2 + r^2)^{1/2} + p \sin\varphi' - c\cos\varphi' \tag{77}$$

whereas the yield condition is expressed by (1). The constitutive equations now have the form

$$\dot{\epsilon}_q = \dot{\lambda}\frac{q}{R} + 2\vartheta_1 \overset{\triangledown}{q}, \quad \dot{\epsilon}_r = \dot{\lambda}\frac{r}{R} + 2\vartheta_1 \overset{\triangledown}{r}, \qquad \dot{\epsilon}_p = \dot{\lambda}\sin\varphi' + \frac{2\vartheta_1(\vartheta_1 - 3\vartheta_2)}{\vartheta_1 - \vartheta_2}\overset{\triangledown}{p} . \tag{78}$$

Denoting

$$k_1 = 2\vartheta_1 , \qquad k_2 = \frac{2\vartheta_1(\vartheta_1 - 3\vartheta_2)}{\vartheta_1 - \vartheta_2} , \tag{79}$$

the relations (77) are expressed as follows

$$\dot{\epsilon}_q = \dot{\lambda}\frac{q}{R} + k_1 \overset{\triangledown}{q} , \quad \dot{\epsilon}_r = \dot{\lambda}\frac{r}{R} + k_1 \overset{\triangledown}{r}, \qquad \dot{\epsilon}_p = \dot{\lambda}\sin\varphi' + k_2 \overset{\triangledown}{p} . \tag{80}$$

These relations can easily be interpreted in the space (q, r, p). The stress rate vector $\overset{\triangledown}{\underset{\sim}{t}}(\overset{\triangledown}{q}, \overset{\triangledown}{r}, \overset{\triangledown}{p})$ is tangential to the cone representing the yield surface. The plastic strain rate is composed of two terms: one normal to the potential surface $g(\underset{\sim}{\sigma}) = 0$, the other uniquely related to $\overset{\triangledown}{t}$.

Using the definitions of the stress rate $\overset{\triangledown}{\sigma}_{ij}$, we have for the plane case

$$
(81)\qquad
\begin{aligned}
&\overset{\triangledown}{\sigma}_{11} = \overset{\triangledown}{\sigma}_x = \dot{\sigma}_x - 2\tau_{xy}\,\omega_{xy} && \overset{\triangledown}{\tau}_{yz} = \overset{\triangledown}{\tau}_{zx} = 0\,,\\
&\overset{\triangledown}{\sigma}_{22} = \overset{\triangledown}{\sigma}_y = \dot{\sigma}_y + 2\tau_{xy}\,\omega_{xy} && \overset{\triangledown}{\sigma}_z = \dot{\sigma}_z\\
&\overset{\triangledown}{\sigma}_{12} = \overset{\triangledown}{\tau}_{xy} = \dot{\tau}_{xy} + (\sigma_x - \sigma_y)\,\omega_{xy}\,,
\end{aligned}
$$

where $\omega_{xy} = \omega_{12} = -\,\omega_{21} = 1/2(\partial u/\partial y - \partial v/\partial x)$. From (81) it follows that

$$
(82)\qquad \overset{\triangledown}{q} = \frac{1}{2}(\overset{\triangledown}{\sigma}_x - \overset{\triangledown}{\sigma}_y) = \dot{q} - 2r\omega_{xy}\,,\quad \overset{\triangledown}{r} = \overset{\triangledown}{\tau}_{xy} = \dot{r} + 2q\omega_{xy}\,,\quad \overset{\triangledown}{p} = \frac{1}{2}(\overset{\triangledown}{\sigma}_x + \overset{\triangledown}{\sigma}_y) = \dot{p}
$$

and since $q = R\cos 2\psi$, $r = R\sin 2\psi$, we can write

$$
(83)\qquad \overset{\triangledown}{q} = \dot{R}\cos 2\psi - 2R\sin 2\psi(\dot{\psi} + \omega_{xy}),\quad \overset{\triangledown}{r} = \dot{R}\sin 2\psi + 2R\cos 2\psi(\dot{\psi} + \omega_{xy}),\quad \overset{\triangledown}{p} = p.
$$

Using the yield condition $R(p) = c\cos\varphi - p\sin\varphi$ and (83), the constitutive equations (80) are expressed in the form

$$
(84)\qquad
\begin{aligned}
\dot{\epsilon}_q &= \dot{\lambda}\cos 2\psi - k_1[\dot{p}\sin\varphi\cos 2\psi + 2R\sin 2\psi(\dot{\psi} + \omega_{xy})]\,,\\
\dot{\epsilon}_r &= \dot{\lambda}\sin 2\psi - k_1[\dot{p}\sin\varphi\sin 2\psi - 2R\cos 2\psi(\dot{\psi} + \omega_{xy})],\\
\dot{\epsilon}_p &= \dot{\lambda}\sin\varphi' + k_2\dot{p}\,.
\end{aligned}
$$

Eliminating $\dot{\lambda}$ from (84), the following system of two differential equations is obtained

$$
(85)\qquad
\begin{aligned}
&[\cos 2\psi - k_1 R]\frac{\partial u}{\partial y} + \sin 2\psi\frac{\partial v}{\partial y} - 2\psi\frac{\partial u}{\partial x} + [\cos 2\psi \quad k_1 R]\frac{\partial v}{\partial x} = 2k_1 R\dot{\psi}\,,\\
&-k_1 R\sin\varphi'\sin 2\psi\frac{\partial u}{\partial y} + (\cos 2\psi + \sin\varphi')\frac{\partial v}{\partial y} + (\cos 2\psi - \sin\varphi')\frac{\partial u}{\partial x} +\\
&+ k_1 R\sin\varphi'\sin 2\psi\frac{\partial v}{\partial x} = (k_1\sin\varphi'\sin\varphi + k_2)\cos 2\psi\,\dot{p} +\\
&+ 2k_1 R\sin\varphi'\sin 2\psi\,\dot{\psi}\,.
\end{aligned}
$$

This is a quasi-linear set with respect to derivatives of u and v and it is hyperbolic provided

$$-\frac{1}{R} \leqslant k_1 \leqslant \frac{1}{R} . \tag{86}$$

Similarly as previously, the characteristic directions are defined by the angles ψ'_1 and ψ'_2 between the principal stress trajectory and the lines α and β . We have

$$\mathrm{tg}\psi'_1 = \frac{\cos\varphi' \sqrt{1 - k_1^2 R^2}}{(1 - \sin\varphi')(1 + k_1 R)}, \qquad \mathrm{tg}\psi'_2 = \frac{\cos\varphi' \sqrt{1 - k_1^2 R^2}}{(1 - \sin\varphi')(1 + k_1 R)} . \tag{87}$$

It is seen that the angles ψ'_1 and ψ'_2 depend on the parameter k_1 but not on k_2. Let us now investigate the case when the velocity and stress characteristics coincide. Requiring $\psi'_1 = \pi/4 + \varphi/2$ or $\psi'_2 = -(\pi/4 + \varphi/2)$ from (87) we find

$$k_1 = -\frac{1}{R} \frac{\sin\varphi - \sin\varphi'}{1 - \sin\varphi \sin\varphi'} . \tag{88}$$

The characteristic equations in the general case are

$$\begin{aligned}
&\frac{\partial v_1}{\partial s_\alpha} + \cos 2\mu_k \frac{\partial v_2}{\partial s_\alpha} - v_2 \sin 2\mu_k \frac{\partial \beta_2}{\partial s_\alpha} - \mathrm{tg}\zeta \cos\varphi' \Big(\frac{\partial v_2}{\partial s_\alpha} \sin 2\mu_k + v_2 \cos 2\mu_k \frac{\partial \beta_2}{\partial s_\alpha} + \\
&+ v_1 \frac{\partial \beta_1}{\partial s_\alpha}\Big) = -\frac{1}{2}(k_1 \sin\varphi' \sin\varphi + k_2)\dot{p} + \mathrm{tg}\varphi \cos\varphi' \dot{\psi} = 0 , \\
&\frac{\partial v_2}{\partial s_\beta} + \cos 2\mu_k \frac{\partial v_1}{\partial s_\beta} + v_1 \sin 2\mu_k \frac{\partial \beta_1}{\partial s_\beta} + \mathrm{tg}\zeta \cos\varphi' \Big(-\frac{\partial v_1}{\partial s_\beta} \sin 2\mu_k + \\
&+ v_1 \cos 2\mu_k \frac{\partial \beta_1}{\partial s_\beta} + v_2 \frac{\partial \beta_2}{\partial s_\beta}\Big) = -\frac{1}{2}(k_1 \sin\varphi' \sin\varphi' + k_2)\dot{p} - \mathrm{tg}\zeta \cos\varphi' \dot{\psi} = 0 ,
\end{aligned} \tag{89}$$

where v_1, v_2 are velocity components referred to s_α and s_β directions and

$$\sin\zeta = k_1 R , \quad \mathrm{tg}\mu_k = \frac{\cos\varphi' \cos\zeta}{(1 - \sin\varphi')(1 + \sin\zeta)} , \quad \mu_k = \psi'_1 = -\psi'_2 . \tag{90}$$

Consider now the particular case

$$\varphi_1' = 0, \quad k_2 = 0, \quad k_1 = -\frac{\sin\varphi}{R} \tag{91}$$

that is incompressible deformation with coincident stress and velocity characteristics. Now

$$\sin\zeta = -\sin\varphi, \quad \operatorname{tg}\mu_k = \frac{\cos\varphi}{1-\sin\varphi}, \quad \mu_k = \pi/4 + \frac{\varphi}{2} \tag{92}$$

and (88) become

$$\begin{aligned} \beta_1 = \psi - \left(\frac{\pi}{4} + \frac{\varphi}{2}\right): \quad & \cos\varphi \frac{\partial v_1}{\partial s_\alpha} - (v_2 - v_1 \sin\varphi)\frac{\partial \psi}{\partial s_\alpha} - \sin\varphi\,\dot{\psi} = 0, \\ \beta_2 = \psi + \left(\frac{\pi}{4} + \frac{\varphi}{2}\right): \quad & \cos\varphi \frac{\partial v_2}{\partial s_\beta} - (v_1 - v_2 \sin\varphi)\frac{\partial \psi}{\partial s_\beta} + \sin\varphi\,\dot{\psi} = 0. \end{aligned} \tag{93}$$

where $\dot{\psi} = \partial\psi/\partial t + v_1\, \partial\psi/\partial s_\alpha + v_2\, \partial\psi/\partial s_\beta$.

The characteristic relations (89) can also be given a simple interpretation by using the formulae for the rate of extension and rotation of characteristic lines. We have

$$\begin{aligned} \dot{\epsilon}_\alpha &= \frac{\partial v_1}{\partial s_\alpha} + \cos 2\mu_k \frac{\partial v_2}{\partial s_\alpha} - v_2 \sin 2\mu_k \frac{\partial \beta_2}{\partial s_\alpha}, \\ \dot{\epsilon}_\beta &= \frac{\partial v_2}{\partial s_\beta} + \cos 2\mu_k \frac{\partial v_1}{\partial s_\beta} + v_1 \sin 2\mu_k \frac{\partial \beta_1}{\partial s_\beta}, \\ \dot{\omega}_\alpha &= \frac{\partial v_2}{\partial s_\alpha} \sin 2\mu_k + v_2 \cos 2\mu_k \frac{\partial \beta_2}{\partial s_\alpha} + v_1 \frac{\partial \beta_1}{\partial s_\alpha}, \\ \dot{\omega}_\beta &= -\frac{\partial v_1}{\partial s_\beta} \sin 2\mu_k + v_1 \cos 2\mu_k \frac{\partial \beta_1}{\partial s_\beta} + v_2 \frac{\partial \beta_2}{\partial s_\beta}. \end{aligned} \tag{94}$$

Using these relations, the characteristic equations (89) can be expressed in the form

$$\begin{aligned} &\dot{\epsilon}_\alpha + (\omega_\alpha - \dot{\psi})\operatorname{tg}\zeta\cos\varphi' - \tfrac{1}{2}(k_1 \sin\varphi' \sin\varphi + k_2)\dot{p} = 0, \\ &\dot{\epsilon}_\beta + (\dot{\omega}_\beta - \dot{\psi})\operatorname{tg}\zeta\cos\varphi' - \tfrac{1}{2}(k_1 \sin\varphi' \sin\varphi + k_2)\dot{p} = 0, \end{aligned} \tag{95}$$

and the equations (93) are reduced to simple relations expressing the rate of extension of characteristic lines in terms of their rate of rotation and the rotation rate of principal stress axes.

$$\dot{\epsilon}_\alpha + (\dot{\omega}_\alpha - \dot{\psi})\,\mathrm{tg}\varphi = 0\,,$$
$$\dot{\epsilon}_\beta - (\dot{\omega}_\alpha - \dot{\psi})\,\mathrm{tg}\varphi = 0\,. \tag{96}$$

It can easily be checked that (93) or (96) are equivalent to characteristic relations derived by Spencer (13) and later by Mandl (10) who started from the assumption that plastic flow is a superposition of two shearing mechanisms along the stress characteristics whose orientation varies due to stress redistribution in a moving material particle. However, our analysis is more general since it covers both dilating flow and elasticity effect. In fact, k_1 in (88) may represent both elastic and plastic properties, whereas the angle φ' defines dilatancy of the material.

The presented analysis may prove useful in experimental verification of the stress rate effect. Assume that the material is rigid-plastic, for instance, a set of grains whose elasticity may be ignored. Assume flow to be incompressible and let the flow rule have the form

$$\dot{\epsilon}^P_{ij} = \dot{\lambda}\, s_{ij} + k_1 \overset{\triangledown}{\sigma}_{ij} = \dot{\epsilon}^P_{ij(1)} + \dot{\epsilon}^P_{ij(2)} \tag{97}$$

where $\overset{\triangledown}{\sigma}_{ij}$ is represented by a vector, tangential to the yield surface. In the plane case, we have

$$\dot{\epsilon}_q = \dot{\lambda}\frac{q}{R} + k_1\overset{\triangledown}{q}\,, \quad \dot{\epsilon}_r = \dot{\lambda}\frac{r}{R} + k_1\overset{\triangledown}{r}\,, \quad \dot{\epsilon}_p = 0 \tag{98}$$

and the condition of coincidence of characteristics requires $k_1 = -\sin\varphi/R$. In the plane (q, r), the total strain rate is represented by two vectors: $\dot{\epsilon}_1$ normal to the yield locus and $\dot{\underset{\sim}{\epsilon}}_2$ tangential to the yield surface with opposite direction to that of $\overset{\triangledown}{\underset{\sim}{\sigma}}$, Fig. 4. Thus,

$$A = \overset{\triangledown}{\underset{\sim}{\sigma}} \cdot \dot{\underset{\sim}{\epsilon}} = \dot{\underset{\sim}{\sigma}} \cdot \dot{\underset{\sim}{\epsilon}} = \dot{\underset{\sim}{\sigma}} \cdot \dot{\underset{\sim}{\epsilon}}_2 = -\frac{\sin\varphi}{R}\,(\overset{\triangledown}{q}^2 + \overset{\triangledown}{r}^2) < 0\,. \tag{99}$$

In other words, if the stress and velocity characteristics are to coincide in a rigid-plastic material, the scalar product of rates of stress and strain should be negative. Let us note that elastic constants η_1 , η_2 are positive and this product for elastic strain rates is always positive. Equation (99) can also be used to direct experimental verification of the hypothesis on combined shearing mechanism in plane flow.

Finally, let us note that using the representation (25), we find that the flow rate (80) can be reduced to (25) provided the parameters $\dot{a}$, $\dot{b}$ and $\dot{\lambda}_3$ depend on the stress rate, namely

$$\dot{a} = \dot{\lambda} + \frac{k_1}{R}(q\overset{\nabla}{q} + r\overset{\nabla}{r}), \quad \dot{b} = \frac{k_1}{R}(r\overset{\nabla}{q} - q\overset{\nabla}{r}),$$

(100)

$$\dot{\lambda}_3 \sin\alpha = \dot{\lambda}\sin\varphi' + k_2\overset{\nabla}{p},$$

and the expressions $q\overset{\nabla}{q} + r\overset{\nabla}{r}$ and $r\overset{\nabla}{q} - q\overset{\nabla}{r}$ are invariants with respect to rotation of the coordinate axes x, y within the plane of flow.

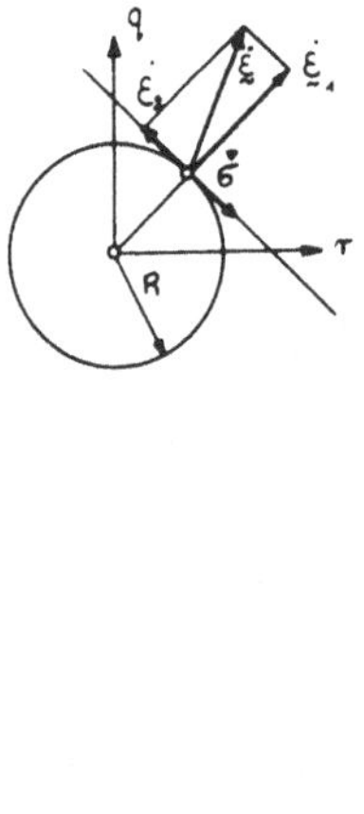

Fig. 4

4. Solution of Some Boundary-value Problems

In this section, we shall discuss application of some of discussed flow rules to solution of boundary-value problems. It will be shown that when stress and velocity characteristics do not coincide, the difficulty in solving the problem increases considerably. Further, we shall discuss some experimental data aimed at verification of theoretical models.

Let us first write down the characteristic relations for several models that will be applied in solving the boundary-value problems

(i) The associated flow rule

The plastic strain-rate is normal to the yield surface and the axes of stress and strain-rate tensors coincide. The velocity equations are given by (60)

$$\frac{\partial v_1}{\partial s_\alpha} - \sin\varphi \frac{\partial v_2}{\partial s_\alpha} - \cos\varphi \frac{\partial \psi}{\partial s_\alpha} = 0 , \quad \frac{\partial v_2}{\partial s_\beta} - \sin\varphi \frac{\partial v_1}{\partial s_\beta} + v_2 \cos\varphi \frac{\partial \psi}{\partial s_\beta} = 0 , \qquad (101)$$

and the characteristic equations are defined as follows

$$\frac{dy}{dx} = \mathrm{tg}[\psi \pm (\frac{\pi}{4} + \frac{\varphi}{2})] . \qquad (102)$$

Now the stress and velocity characteristics coincide and the relations (101) express the fact that the elongation rates along characteristic lines are zero $\dot{\epsilon}_\alpha = \dot{\epsilon}_\beta = 0$. The relations (101) can also be expressed in terms of normal projections v_α^t , v_β^t of the velocity vector on the lines α and β

$$\frac{\partial v_\alpha^t}{\partial s_\alpha} - (v_\alpha^t \,\mathrm{tg}\varphi + v_\beta^t \sec\varphi) \frac{\partial \psi}{\partial s_\alpha} = 0 ,$$
$$\frac{\partial v_\beta^t}{\partial s_\beta} + (v_\alpha^t \sec\varphi + v_\alpha^t \,\mathrm{tg}\varphi) \frac{\partial \psi}{\partial s_\beta} = 0 . \qquad (103)$$

Along the line of strong discontinuity which coincides with the one of stress characteristics, the jumps in tangential and normal velocity follow the relations

$$\Delta v^t = \Delta v_o^t \exp \mathrm{tg}\varphi(\psi - \psi_o) , \qquad \Delta v^n = \Delta v^t \,\mathrm{tg}\varphi , \qquad (104)$$

which can be derived from (101) or (103). Here Δv^t and Δv^n denote jumps in tangential and normal velocities to the discontinuity lines. The second relation (104) implies that the velocity discontinuity vector is inclined at the angle to the line of discontinuity.

Fig. 5a shows the characteristic mesh on the physical plane and the mapping on the hodograph plane is shown in Fig. 5b. Since the characteristic relations express the fact that the lines α and β are inextensible, the mapping on the hodograph plane is represented by an orthogonal net to the velocity characteristics. Fig. 5c shows the line of discontinuity and Fig. 5d presents its mapping on the hodograph plane. This orthogonality property of the hodograph suffices to determine the velocity field if proper boundary conditions are specified. For instance, if velocities at the points 2 and 3 are given, the velocity at the point 4 is determined by drawing orthogonal lines to 2-4 and 1-4 on the hodograph plane and determining the position 4.

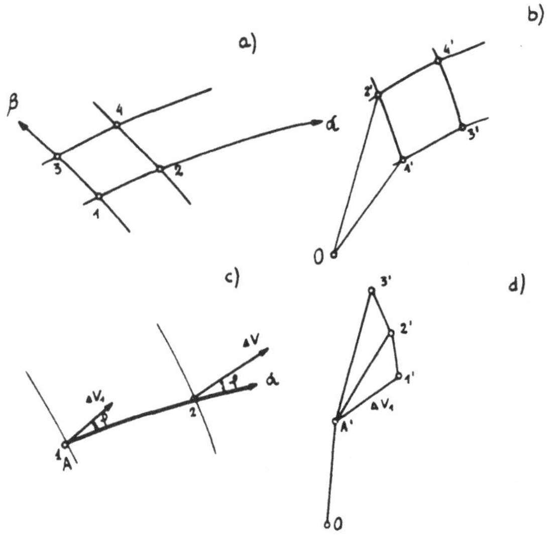

Fig. 5

(ii) Incompressible flow: coaxial flow rule

Now, the characteristic relations are

$$\frac{\partial v_\alpha^t}{\partial s_\alpha} - v_\beta^t \frac{\partial \psi}{\partial s_\alpha} = 0\,, \qquad \frac{\partial v_\beta^t}{\partial s_\beta} + v_\alpha^t \frac{\partial \psi}{\partial s_\beta} = 0 \tag{105}$$

and the velocity characteristics are orthogonal, that is

$$\left(\frac{dy}{dx}\right)_{\alpha,\beta} = \mathrm{tg}\left(\psi \pm \frac{\pi}{4}\right). \tag{106}$$

The velocity characteristics do not coincide with the stress characteristics: the α -line deviates from the stress characteristics s_1 by the angle $1/2\,\varphi$ and the β -line forms the angle $-1/2\varphi$ with the line s_2.

The velocity equations (105) are identical to Geiringer equations used in studying plastic flow of metals and the hodograph construction follows the orthogonality rule, see Fig. 5. Now, however, an additional difficulty arises since the mesh of velocity characteristics should be constructed once the field of stress characteristics has been found.

(iii) Incompressible flow: non-coaxial flow rule

Only two limiting cases will be considered which correspond to Geniev hypothesis. Assume that the principal stress and strain rate axes make the angles $+\ 1/2\varphi$ or $-\ 1/2\varphi$. The kinematic equations now are

$$\frac{\partial u}{\partial x} + \frac{\partial v}{\partial y} = 0\,,$$

$$\left(\frac{\partial u}{\partial x} - \frac{\partial v}{\partial y}\right)\sin(2\psi \pm \varphi) - \left(\frac{\partial v}{\partial x} + \frac{\partial u}{\partial y}\right)\cos(2\psi \pm \varphi) = 0\,, \tag{107}$$

and the velocity characteristics are

$$\frac{dy}{dx} = \mathrm{tg}\left[\psi \mp \left(\frac{\pi}{4} + \frac{\varphi}{2}\right)\right], \qquad (\alpha', \beta')$$

or

$$\frac{dy}{dx} = tg[\psi \mp (\frac{\pi}{4} - \frac{\varphi}{2})]. \qquad (\alpha'', \beta'') \tag{108}$$

It is seen that there there are two sets of characteristics depending on whether the deviation $+\ 1/2\varphi$ or $-1/2\varphi$ occurs. The decision which set should be used in any particular problem depends on the problem considered. One of possible criterions for this decision is the positive rate of dissipation within the plastic domain. Figure 6 shows the relative orientation of stress and velocity characteristics for all three flow rules.

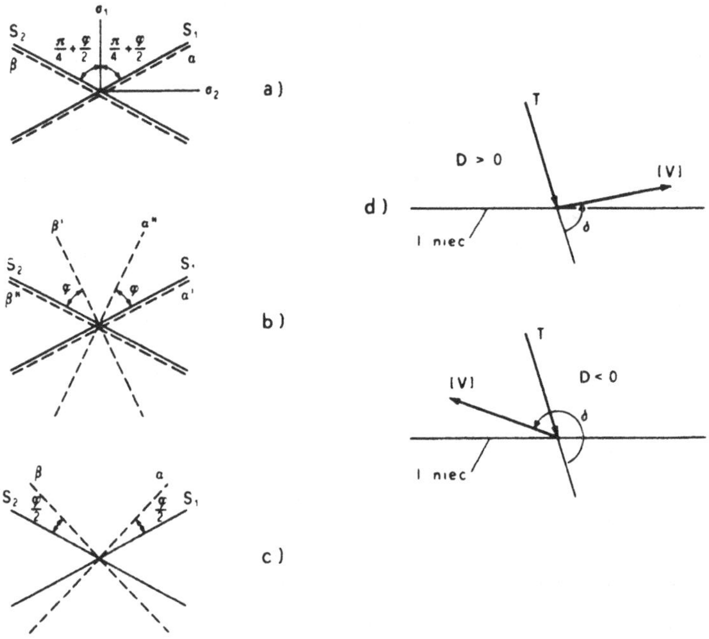

Fig. 6

(iv) The condition of positive dissipation power

In formulating flow rules, we assume that the plastic strain-rate vector is directed in the exterior domain beyond the elastic region enclosed by the yield surface. In other words, we assume that $\dot{\lambda}_1, \dot{\lambda}_2, \dot{\lambda}_3$ in (18) are positive. This requirement follows from the fact that any yielding mechanism must be associated with positive rate of energy dissipation. Since our frictional material is regarded as a purely dissipative system, we should require that

$$D = \sigma_{ij}\dot{\epsilon}_{ij} = p\dot{\epsilon}_p + q\dot{\epsilon}_q + r\dot{\epsilon}_r > 0 \,. \tag{109}$$

For the coaxial flow rule (25), we obtain

$$D = p\dot{\lambda}_3 \sin\alpha + q\dot{a}\cos 2\psi + r\dot{a}\sin 2\psi = p\dot{\lambda}_3 \sin\alpha + R\dot{a} \,. \tag{110}$$

Since $q = R\cos 2\psi$, $r = R\sin 2\psi$, $R = c\cos\varphi - p\sin\varphi$, Eq. (105) can be written in the form

$$D = \dot{\lambda}\,[c\cos\varphi + p(\sin\varphi' - \sin\varphi)] \tag{111}$$

where $\dot{\lambda} = \dot{\lambda}_3 \cos\alpha$, $\sin\varphi' = \mathrm{tg}\,\alpha$. We see that the dissipation function D depends on the mean stress p. To express $\dot{\lambda}$ in terms of the principal strain rates, we have

$$\left.\begin{matrix}\dot{\epsilon}_1\\ \dot{\epsilon}_2\end{matrix}\right\} = \frac{1}{2}\dot{\epsilon}_p \pm \frac{1}{2}(\dot{\epsilon}_q^2 + \dot{\epsilon}_r^2)^{1/2} = \frac{1}{2}\dot{\epsilon}_p \pm \frac{1}{2}\dot{a} \tag{112}$$

and $|\dot{\epsilon}_1 - \dot{\epsilon}_2| = \dot{\lambda} = \dot{a}$. Thus

$$D = |\dot{\epsilon}_1 - \dot{\epsilon}_2|\,[c\cos\varphi + p(\sin\varphi' - \sin\varphi)\,] \,. \tag{113}$$

Similarly, for the non-coaxial flow rules (42) and (43), we find

$$D = \dot{\lambda}R\cos\varphi = |\dot{\epsilon}_1 - \dot{\epsilon}_2|\,(c\cos\varphi - p\sin\varphi)\cos\varphi \,. \tag{114}$$

At the vertex A, the dissipation function equals $D_A = c \operatorname{ctg} \varphi\ (\dot{\epsilon}_1 + \dot{\epsilon}_2)$. It is seen that the dissipation function is expressed in terms of strain rates only for the associated flow rule, $\varphi = \varphi'$. In other cases, the mean stress p should be known in order to determine the rate of dissipation.

To verify whether the rate of dissipation is positive, both stress and strain rates should be determined within the plastic region and the inequality (108) should be checked at each element. Similarly, along the discontinuity line, the direction of shearing stress should be the same as that of the velocity jump, see Fig. 6d.

4.1. Example 1: Motion of a Vertical Plane into a Soil Occupying One Quadrant

Consider a simple boundary-value problem: a rigid plate AE moves horizontally with the prescribed velocity v_c into the soil which is assumed to be cohesionless and satisfying the Coulomb yield condition. The angle of friction between the soil and the plate equals φ_ω. This is the classical problem of passive earth pressure and the static solution can easily be constructed.

For a weightless soil, in order to avoid trivial solution with vanishing stresses, assume that along AB the uniform vertical surface loading is applied, Fig. 7. The

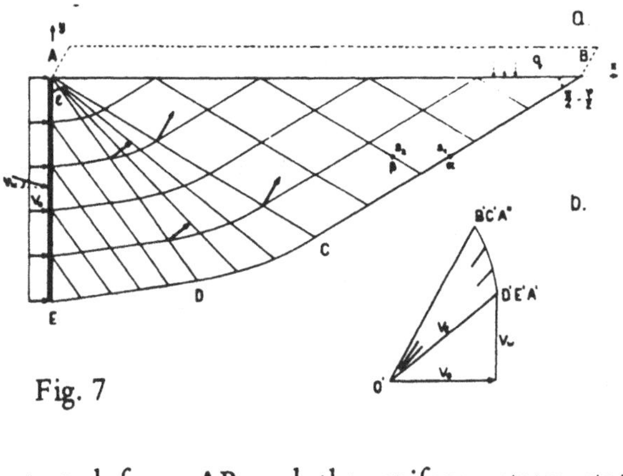

Fig. 7

stress solution is started from AB and the uniform stress state occurs in ABC, bounded by stress characteristics AC and CB. The solution within a centered fan ACD is obtained by assuming that A is singularity point with a degenerating family of s_1-lines. Now, s_2-lines are logarithmic spirals $r = r_0 \exp(\theta \operatorname{tg} \varphi)$ referred to the polar coordinated system (r, θ) with the center at A. The angle ϵ determining the inclination of s_2-lines to the plate is found from the friction condition $\varphi_\omega = \operatorname{arctg} \mu$. Upon using the Coulomb yield condition, we obtain

$$\epsilon = \pi/4 + \frac{1}{2}\varphi - \frac{1}{2}\varphi_w - \frac{1}{2} \arcsin \frac{\sin \varphi_w}{\sin \varphi} . \tag{115}$$

The uniform stress state within ADE now satisfies the boundary conditions on AE. Thus within ABC and ADE both s_1 and s_2 are straight lines. The characteristic network in Fig. 7 corresponds to $\varphi = 30°$, $\varphi_\omega = 15°$. When φ_ω increases, the angle ϵ diminishes and the region ADE vanishes when $\varphi_\omega = \varphi$.

Now, let us pass to determination of the velocity field. Consider first the associated flow rule. Now the velocity characteristics coincide with the stress characteristics and the lines EDCB and EA are assumed as the strong discontinuity lines with DCB separating the rigid region from the flow region. We start from the point E and decompose v_0 into v_e and v_w, Fig. 7b, where v_w denotes the velocity of gliding along AE and v_e is the velocity along the discontinuity line EDCB. The mapping on the hodograph plane is now easily performed: first we determine variation of the velocity jump along the discontinuity line and next velocity in each region. Note that ADE and ACB move like rigid blocks whereas the region ACD deforms. The dotted line shows the deformed shape of AB during the incipient plastic flow. The next step would require construction of a new characteristic mesh for the stress equations. It can be checked that this velocity field satisfies the condition of positive dissipation power and is kinematically admissible.

Now, let us consider the Geniew non-coaxial flow rule (107) and assume the active velocity characteristics α' coincide with the s_1-lines. The second family β' is orthogonal to α' and the shape of β'-lines is shown in Fig. 8, where the hodograph construction is also illustrated. Similarly as previously, we start from the point E and decompose the velocity v_0 into v_e and v_w tangential to the line of discontinuity and the plate. Along EHDCB the velocity is constant and its mapping on the hodograph plane is shown by D′B′. In AHE the problem is of mixed type: the known velocity on EH and normal velocity v_0 determine the solution within AHE. Similarly, the solution within ABCD is determined starting from the condition on AH and HDCB. It is easy to see that this velocity field is different from that corresponding to the associated flow rule. The whole region AFDE moves as a rigid body with the velocity v_e and the region GBC is sheared along the lines α'. The shape of the deformed surface is shown by the dotted line.

Checking whether the dissipation power within the flow region is positive, it turns out that within GBC there is $D < 0$ since the major stress $\sigma_x < 0$ and

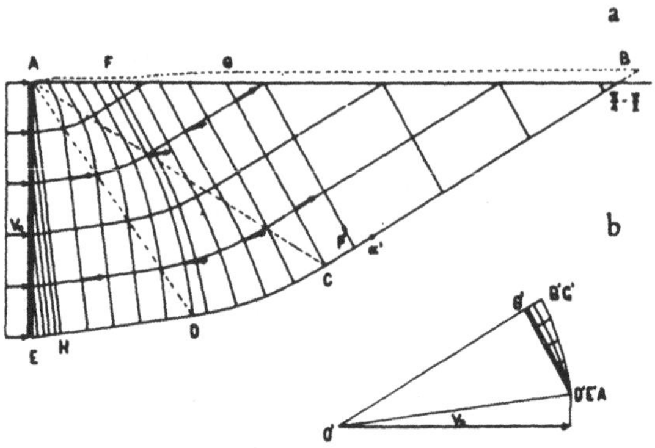

Fig. 8

$\dot{\epsilon}_x > 0$ within this region. Thus the kinematic solution is inadmissible in this region since it corresponds to the strain rate vector pointing to the interior of the yield surface.

Now, let us discuss the second limiting case when the lines s_2 and β'' coincide. Now the discontinuity line EHGF separating rigid and plastic regions coincides with one of orthogonal velocity characteristics α''. The net of velocity characteristics and the mapping on the hodograph plane is shown in Fig. 9. The dissipation power is everywhere positive within the plastic region.

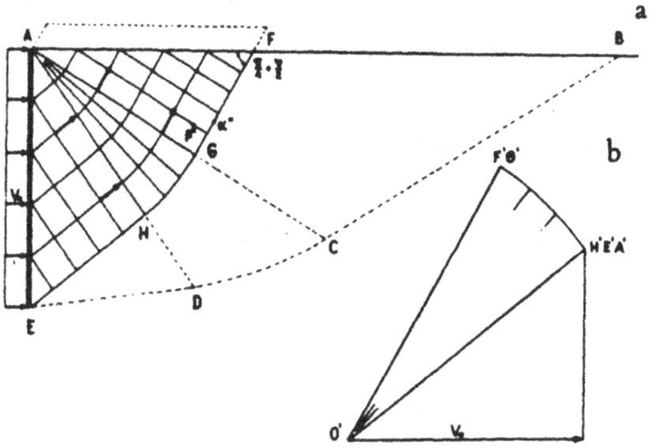

Fig. 9

Consider finally the coaxial and incompressible flow rule (104) and (105). The velocity characteristics are now symmetrically situated with respect to stress characteristics and make angles $1/2\,\varphi$ with the lines s_1 and s_2. Within the regions AHE and AFG the lines α and β are straight whereas within AGH the α-lines are logarithmic spirals $r = r_0 \exp(\theta \operatorname{tg}\varphi/2)$ and the β-lines are spirals $r = r_0 \exp(-\theta \operatorname{tg}\varphi/2)$, see Fig. 10. There is no singularity at the point A, similarly as in Fig. 8. The hodograph is shown in Fig. 10. It turns out that the dissipation power is negative within LFG and the constructed velocity field is not admissible.

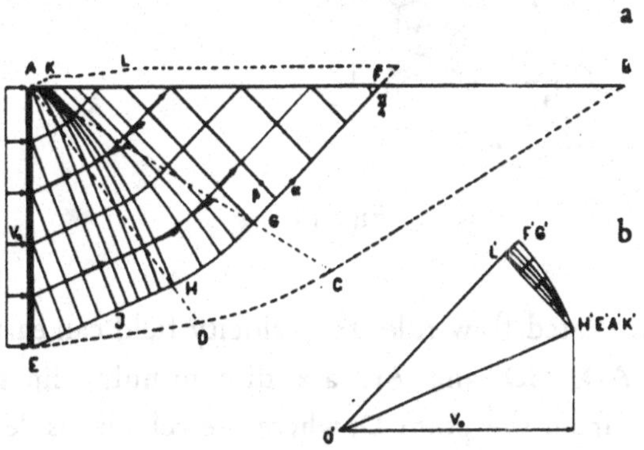

Fig. 10

It is seen therefore that only for two cases the determined velocity field can be regarded as kinematically admissible. This difficulty in solving velocity boundary-value problems is typical for cases when non-associated flow rules are applied which require the different nets of stress and velocity characteristics.

4.2. Example 2: Steady Flow Through Flat-bottom Hopper

Consider a quasi-static flow of a granular material through a vertical plane hopper with central opening in a flat bottom, Fig. 11. For a high hopper it can be assumed that the material flows with the uniform velocity v_0 at cross-sections sufficiently distant from the opening. Our aim is to determine the velocity field near the outlet where steady plastic flow occurs. Neglect gravity forces and assume that flow is induced by downward acting pressure q_1, whereas the upward reaction from a belt feeder equals q_2. Let the static solution be represented by the characteristic net shown in Fig. 11 for $\varphi = 30°$. Within the region ABC stress is uniform and changes within the region ACD with the β -lines centered at A. The friction between the wall and the material is neglected.

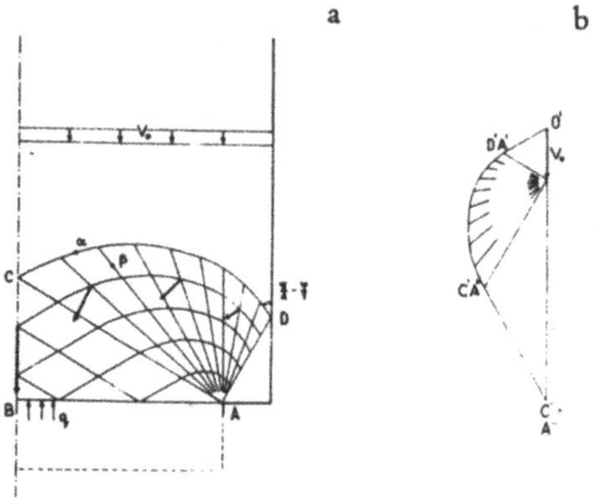

Fig. 11

Assuming the associated flow rule, the velocity field can easily be constructed, Fig. 11. The lines AD, CD and AC are discontinuity lines. The hodograph construction is started from the point D, where the velocity is decomposed into the velocity along DA and the relative velocity along DC. The condition that the velocity below C should be directed along CB enables determining the discontinuity along CA. The region CBA moves downward with a uniform velocity several times exceeding the velocity v_0 because of excessive dilatation of the material predicted by the associated flow rule. This solution, although formally correct, does not seem to be realistic since in actuality the plastic dilatancy is much less and the incompressibility assumption is much closer to reality.

Consider now the Geniev model for which the lines β'' and s_2 coincide. The velocity characteristics and the hodograph are shown in Fig. 12. The velocity profile at the outlet is not uniform and the central portion flows with the velocity v_0, whereas along AF the velocity is greater than v_0 and it is inclined to the symmetry axis. This velocity profile is also not realistic since most experimental data indicate that the velocity of flow increases toward the center of the outlet.

Figures 13 and 14 show the characteristic nets and hodographs for the coaxial flow rule. Depending on the angle of internal friction φ two cases are possible which are controlled by the inequality

Fig. 12

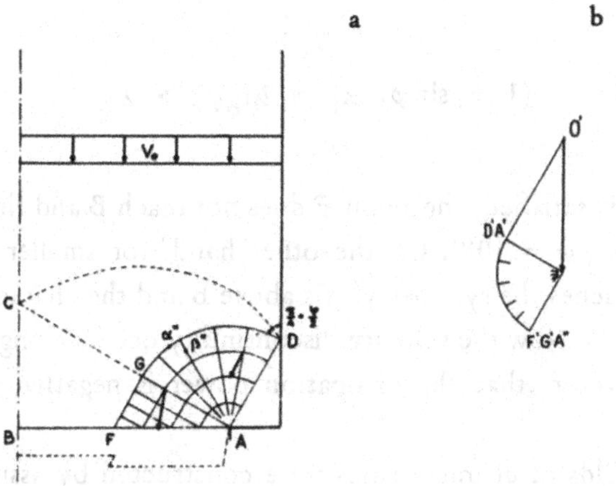

Fig. 13

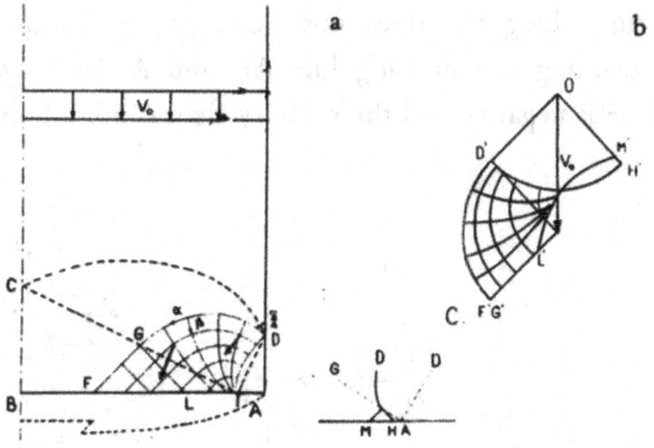

Fig. 14

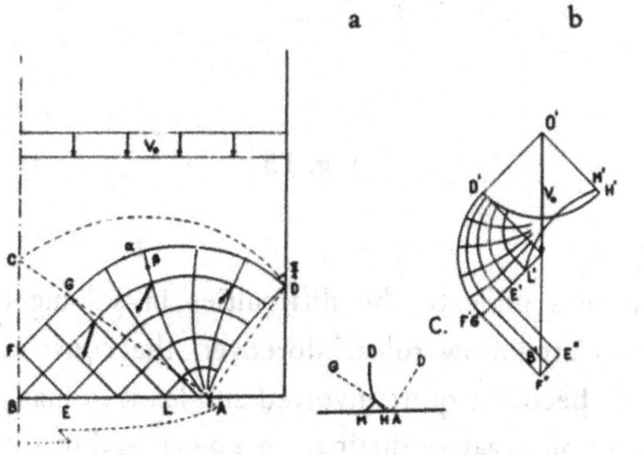

$$(1 + \sin\varphi)\exp(\pi/2\ \mathrm{tg}\varphi) > 2 \tag{116}$$

If this inequality is satisfied, the point F does not reach B and the characteristic net of Fig. 13 applies ($\varphi = 40°$). On the other hand, for smaller values of φ , the α -line from D reaches the symmetry axis above B and the characteristic is shown in Fig. 14 ($\varphi = 20°$). Now the velocity discontinuity occurs along DH, FGD and FE. It turns out, however, that the dissipation power is negative within the domain FGLE.

These two fields of characteristics were constructed by assuming that the line separating rigid and plastic regions starts from D and terminates at H on AB; hence the material portion along HA does not flow. Figure 15 shows the alternative solution by constructing the limiting line AK and A. Here again the dissipation power within FGLE is negative and the velocity field is not admissible in this region.

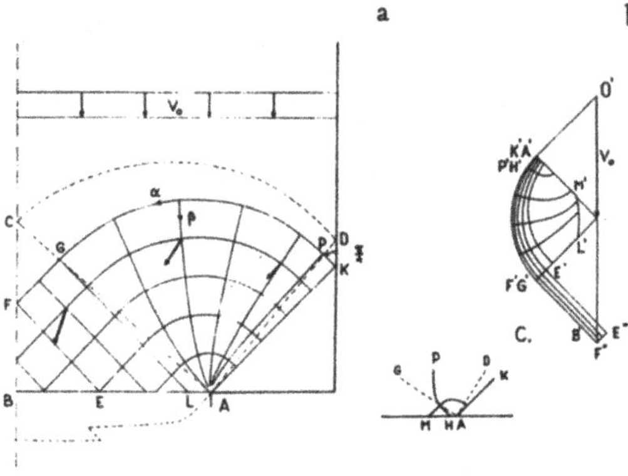

Fig. 15

These two examples indicate the difficulties in solving the boundary value problems for non-associated flow rules. Moreover, the construction of the net of velocity characteristics becomes quite involved and leads in many cases to kinematic inadmissibility because of negative dissipation power within some regions or along discontinuity lines. This difficulty was clearly indicated by Drescher [18]. Further examples of solutions of problems of plane flow can be found in [16-23].

4.3. Experimental Verification

There were numerous tests aimed at verifying the accuracy of theoretical predictions for both static and kinematic fields. For brevity, we shall only present test results reported in [16] and [23]. Figure 16 shows the velocity field observed

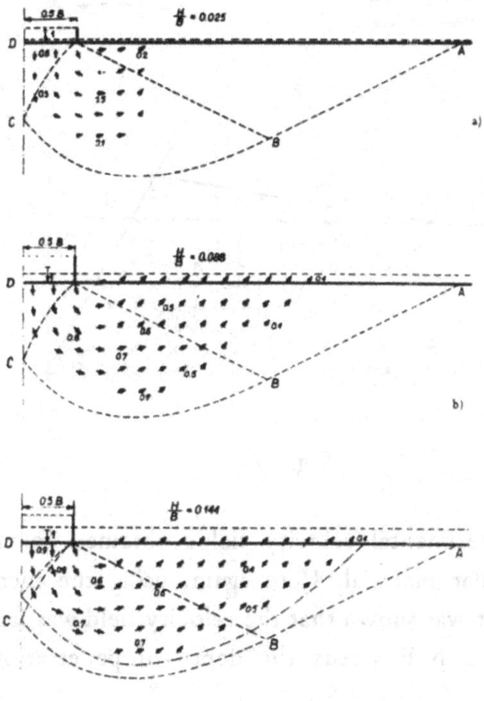

Fig. 16

in the test of a rigid punch identation into a semiplane of granular material (sand). Depending on the depth of penetration the velocity field changes and for H/B = 0.144 the discontinuity line separating plastic and rigid regions is well defined. The prediction of the associated flow rule is not realistic in this case since both the shape of the deforming region and the velocity field are much better described by using the incompressibility and coaxiality assumption. Figure 17 shows

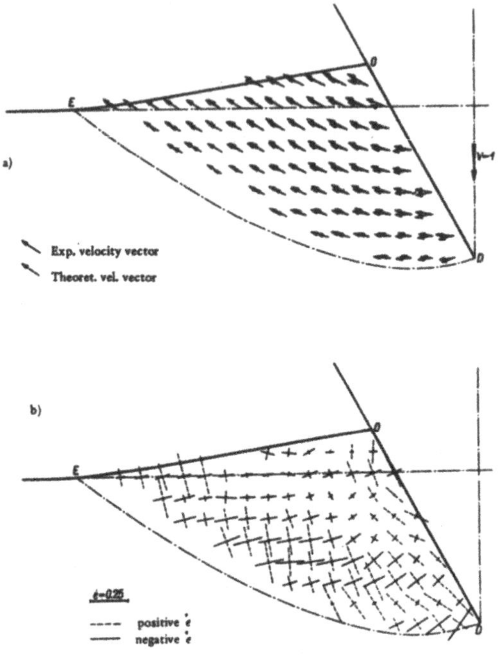

Fig. 17

the theoretical and experimental velocity fields obtained for the case of wedge identation into a granular material. Here again, using the incompressibility and coaxiality assumptions it was shown that the velocity field was fairly well described. Fig. 18 shows the ratio N/F versus the depth of penetration. It is seen that

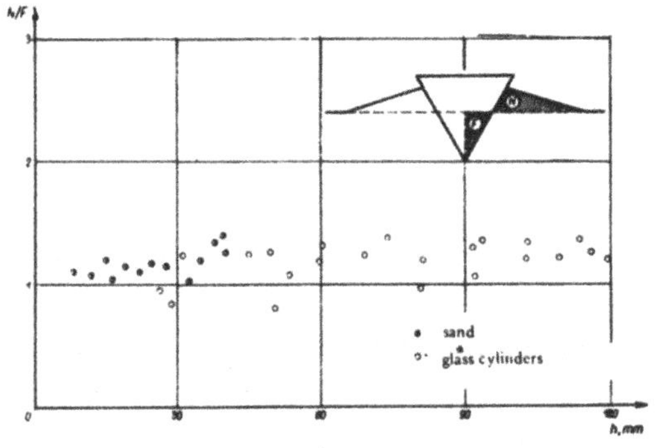

Fig. 18

incompressibility assumption is sufficiently well satisfied although small dilatancy is present. It should be noted however, that for a densely packed material, the incipient plastic flow should be characterized by higher dilatancy and the variable dilatancy model would provide a better description.

REFERENCES

[1] D.C. Drucker and W. Prager, Soil mechanics and plastic analysis or limit design, Quart. Appl. Math., vol. 10, 157-165, 1952.

[2] E. Melan, Zur Plastizitat des raumlichen Kontinuums, Ingenieur Archiv, vol. 9, 1938, 116-126.

[3] R. Hill, The mathematical theory of plasticity, Clarendon Press, 1950.

[4] A. Yu. Ishlinsky, On plane flow of sand, (in Russian), Ukr. Math. Journ. vol. 6, 1954.

[5] A.W. Jenike and R.T. Shield, On the plastic flow of Coulomb solids beyond original failure, J. Appl. Mech. vol. 27, 599-602, 1959.

[6] R.T. Shield, Mixed boundary-value problems in soil mechanics, vol. 11, 1953, 61-75.

[7] S. Takagi, Plane plastic deformation of soils, Journ. Eng. Mech. Div. ASCE, vol. 88, 1962, 101-151.

[8] De Josselin de Jong, Statics and kinematics in the failable zone of a granular material, Delft, 1959.

[9] J. Mandel, Sur les equations d'écoulement des sols ideaux en deformation plane et le concept du double glissement, Journ. Mech. Phys. Solids, vol. 14, 1968, 303-308.

[10] G. Mandl, and F.R. Luque, Fully developed plastic shear flow of granular materials, Géotechnique, vol. 20, 1970, 277-307.

[11] G.A. Geniev, Problems of dynamics of granular media, Akad. Stroit. i Arch. Moscow, 1959.

[12] W.W. Sokolovskii, Statics of soil media, Moscow, 1960.

[13] A.J.M. Spencer, A theory of kinematics of ideal soil under plane strain conditions, Journ. Mech. Phys. Solids, vol. 12, 1964, 337-351.

[14] Z. Mróz, Non-associated flow laws in plasticity, Journ. de Mécanique, vol. 2, 1963, 21-42.

[15] V.N. Nikolaevski, Constitutive equations of plastic flow of deforming granular materials, Prikl. Math and Mekch., 1971, vol. 35, 1070-1082.

[16] A. Drescher, K. Kwaszczynska and Z. Mróz, Statics and kinematics of the granular medium in the case of wedge identation, Arch. Mech. Stos., vol. 19, 1967, 99-113.

[17] Z. Mróz and A. Drescher, Limit plasticity approach to some cases of flow of bulk solids, Journ. Eng. for Industry, vol. 91, 1969, 357-365.

[18] A. Drescher, Some remarks on plane flow of granular media, Arch. Mech. Stos. vol. 24, 1972, 837-848.

[19] A. Drescher and A. Dragon, On kinematics of slopes and embankments, Arch. Inz. Lad. (in Polish), vol.20, 1974.

[20] Z. Mróz and A. Drescher, Foundations of plasticity for granulated materials, (in Polish), Ossolineum, 1972.

[21] W. Szczepinski, On the motion of flat landslides and avalanches treated as a problem of plasticity, Arch. Mech. Stos. vol. 24, 1972, 919-930.

[22] Z. Mróz and K. Kwaszczynska, Axially symmetric plastic flow of soils treated by the graphical methods, Arch. Mech. Inz. Lad. vol. 14, 1968, 27-37.

[23] A. Drescher and A. Bujak, Kinematics of a granular material in the case of punch identation (in Polish) Rozpr. Inz. 1966, vol. 14, 313-325.

[24] J.I. Dais, An isotropic frictional theory for a granular material with or without cohesion, Int. J. Solids a. Struct. a. Struct., 1970, vol. 6, 1185-1191.

PLASTICITE POUR LA MECANIQUE DES SOLS

PAR

JEAN SALENÇON

Ecole Polytechnique et E.N.P.C. (Paris)

AVANT PROPOS

Le but de ces leçons est double: faire en quelque sorte l'état des connaissances en ce qui concerne l'application de la théorie de la Plasticité en Mécanique des Sols, et aussi attirer l'attention sur un certain nombre de lignes de recherche nouvelles et intéressantes aux yeux de l'auteur.

Dans l'ouvrage, Théorie de la Plasticité pour les Applications à la Mécanique des Sols *(ed. Eyrolles, Paris), l'auteur a traité plusieurs des questions évoquées ici, en donnant les démonstrations des théorèmes et les calculs détaillés ainsi qu'une bibliographie assez abondante. Cela nous permettra de ne donner, sur ces questions, que des rappels des principaux résultats et des compléments bibliographiques éventuels, renvoyant à l'ouvrage cité, le lecteur soucieux d'approfondir. Nous insisterons donc sur les idées directrices illustrées par des exemplees typiques d'application.*

PLASTICITE et MECANIQUE DES SOLS

Jusqu'aux travaux de Terzaghi, la Mécanique des Sols était pour l'essentiel la théorie de la poussée des terres établie dans le but d'étudier la stabilité des ouvrages; les travaux de Coulomb, Rankine, Résal et Caquot sont célèbres dans ce domaine.

Avec Terzaghi l'accent a été mis sur le comportement du sol, milieu à trois phases, dans lequel l'eau interstitielle joue un rôle essentiel. On s'est alors préoccupé des problèmes de tassement, de consolidation, laissant de côté le calcul des charges ultimes. Il est apparu beaucoup plus important pour un mécanicien du sol de posséder une longue pratique du matériau que de procéder à des analyses théoriques. C'est sans doute là l'origine d'une boutade désormais classique affirmant que dans Mécanique des Sols il y aurait plus de Sols que de Mécanique.

La théorie de la Plasticité quant à elle, s'est principalement développée à propos des problèmes concernant le métal; en particulier les théorèmes de l'analyse limite, outils essentiels du calcul dans le schéma de comportement rigide-plastique, ont été formulés, il n'y a guère plus de 20 ans, dans le cas du matériau "standard" de Tresca ou de Mises. Le lien avec la Mécanique des Sols se faisait surtout par la théorie des équilibres limites plans (Mandel, Sokolovski) dans le cas du matériau de courbe intrinsèque quelconque.

Il semble maintenant, compte-tenu de l'état actuel de la théorie de la plasticité pour le métal, qui se présente comme un corps de doctrine très cohérent, qu'un rapprochement très fructueux doit être fait sur de nombreux points avec la Mécanique des Sols. Bien des raisonnements employés par le mécanicien des sols –par exemple pour les calculs de capacité portantes– des méthodes utilisées, souvent avec difficultés –calcul à la rupture notamment– se trouvent alors éclairés par référence à leurs homologues dans le cas du métal et des théorèmes ont été établis pour les matériaux plastiques dont le comportement n'obéit pas à la règle de normalité, ce qui semble être le cas des sols.

Ainsi, sans contester que le modèle de comportement plastique classique, et a fortiori rigide-plastique, constitue une schématisation simpliste du comportement réel des sols, il apparaît que les errements usuels du mécanicien des sols dans l'étude de certains problèmes correspondent à l'utilisation de ce schéma, même s'ils paraissent parfois ne reposer que sur des idées intuitives et naturelles comme c'est le cas par exemple avec le calcul à la rupture. Pour ces raisons, un enseignement dégageant la signification des procédés de calculs utilisés en Mécanique des Sols à la lumière de la théorie de la Plasticité nous a paru répondre à un besoin actuel.

MODELE DE COMPORTEMENT PLASTIQUE
et
COMPORTEMENT DES SOLS

1. MODELE CLASSIQUE DE COMPORTEMENT PLASTIQUE

Le modèle de comportement plastique *classique* introduit les deux notions fondamentales de

- critère de plasticité (ou d'écoulement)

et

- règle d'écoulement,

qu'il y a lieu d'envisager séparément.

1.1. Critère de plasticité – Ecrouissage

Le domaine d'élasticité du matériau est caractérisé par une fonction scalaire f du tenseur des contraintes $\underline{\underline{\sigma}}$, appelée fonction de charge (yield-function), telle que:

$$f(\underline{\underline{\sigma}}) < 0 \tag{1.1}$$

corresponde à l'intérieur du domaine d'élasticité.

Le matériau est dit parfaitement plastique (perfectly-plastic) si ce domaine est fixe, indépendant des déformations plastiques; il y a écrouissage positif ou négatif (strain-hardening or strain-softening) si le domaine d'élasticité est entraîné par le point de charge $\underline{\underline{\sigma}}$: dans ce cas on écrira souvent f sous la forme:

$$f(\underline{\underline{\sigma}}, E)$$

où E symbolise l'écrouissage.

Il semble qu'indépendamment de toute autre hypothèse, on peut admettre la *convexité de f* (convexity) par rapport à $\underline{\underline{\sigma}}$ comme une propriété générale pour les matériaux usuels.

Rappelons les critère classiques utilisés pour le métal: Critère de von Mises:

$$f = \frac{1}{2} \operatorname{tr} \underline{\underline{s}}^2 - k^2 \tag{1.2}$$

$\underline{\underline{s}}$ = déviateur de $\underline{\underline{\sigma}}$, k = cission limite;
Critère de Tresca:

$$f = \operatorname*{Sup}_{i,j} (\sigma_i - \sigma_j) - 2k \tag{1.3}$$

σ_i = contraintes principales, k = cission limite.

Dans le cas général, si le matériau étudié est isotrope, f doit être une fonction isotrope de $\underline{\underline{\sigma}}$.

1.2. Règle d'écoulement

Lorsque le point de charge $\underline{\underline{\sigma}}$ est sur la frontière f = 0 du domaine d'élasticité actuel et y demeure, il y a déformation plastique qui se superpose à la déformation élastique. La règle d'écoulement (flow-rule) définit le tenseur vitesse de déformation plastique (plastic strain rate) $\underline{\underline{d}}^p$.

On a de façon générale:

- dans le cas du milieu écrouissable:

$$\begin{cases} \underline{\underline{d}}^p = \underline{\underline{H}}(\underline{\underline{\sigma}}, E) \operatorname{tr}(\dfrac{\partial f}{\partial \underline{\underline{\sigma}}} \cdot \dot{\underline{\underline{\sigma}}}) & \text{si } f(\underline{\underline{\sigma}}, E) = 0 \\ & \text{et } \operatorname{tr}(\dfrac{\partial f}{\partial \underline{\underline{\sigma}}} \cdot \dot{\underline{\underline{\sigma}}}) \geqslant 0 \\ \underline{\underline{d}}^p = 0 & \text{dans les autres cas.} \end{cases} \tag{1.4}$$

- dans le cas du matériau parfaitement plastique:

$$\begin{cases} \underline{\underline{d}}^p = \lambda \underline{\underline{H}}(\underline{\underline{\sigma}}) & \text{si } f(\underline{\underline{\sigma}}) = 0 \quad \text{et } \operatorname{tr}(\dfrac{\partial f}{\partial \underline{\underline{\sigma}}} \cdot \dot{\underline{\underline{\sigma}}}) = 0 \\ \qquad \lambda \geqslant 0 & \\ \underline{\underline{d}}^p = 0 & \text{dans les autres cas.} \end{cases} \tag{1.5}$$

De plus, si le matériau est isotrope $\underline{\underline{d}}^p$ et $\underline{\underline{\sigma}}$ ont un système de directions principales communes.

1.3. Principe du travail maximal – Matériau standard

Le principe du travail maximal posé par Hill exprime que si $\underline{\underline{\sigma}}$ est un tenseur-contrainte plastiquement admissible – $f(\underline{\underline{\sigma}}, E) \leqslant 0$ – auquel $\underline{\underline{d}}^p$ est associé par la loi de comportement plastique, et si $\underline{\underline{\sigma}}^*$ est un autre tenseur contrainte quelconque plastiquement admissible dans le même état – $f(\underline{\underline{\sigma}}^*, E) \leqslant 0$ – on a:

$$\mathrm{tr}[(\underline{\underline{\sigma}} - \underline{\underline{\sigma}}^*)\, \underline{\underline{d}}^p] \geqslant 0 \tag{1.6}$$

Il implique:

- la convexité de f en $\underline{\underline{\sigma}}$
- la normalité de la règle d'écoulement, à savoir:

$$\underline{\underline{d}}^p = \lambda \frac{\partial f(\underline{\underline{\sigma}}, E)}{\partial \underline{\underline{\sigma}}} \quad , \; \lambda \geqslant 0 \tag{1.7}$$

si f est une fonction régulière,

ou $$\underline{\underline{d}}^p \in \lambda\, \partial f(\underline{\underline{\sigma}}, E) \; , \quad \lambda \geqslant 0$$

pour comprendre aussi le cas d'un point conique.

On dit aussi que dans ce cas la règle d'écoulement est associée au critère (associated flow rule), ou encore, suivant Radenkovic, que le matériau est standard.

On remarquera enfin qu'actuellement beaucoup d'auteurs préfèrent plutôt que de parler de principe du travail maximal, expliciter les deux propriétés de CONVEXITE et NORMALITE (l'aspect mathématique prenant ainsi le pas sur la signification mécanique).

Dans le cas particulier du matériau de Misès, on a:

$$\underline{\underline{d}}^p = \lambda \underline{\underline{s}} \quad , \qquad \lambda \geqslant 0 \tag{1.8}$$

et pour le matériau de Tresca en se plaçant dans le repère principal:

$$(1.9) \qquad d_1^p = \lambda\,, \quad d_2^p = 0, \quad d_3^p = -\lambda\,, \quad \lambda \geqslant 0 \text{ si } \sigma_1 > \sigma_2 > \sigma_3$$

et

$$d_1^p = \lambda + \mu, \quad d_2^p = -\mu, \quad d_3^p = -\lambda, \quad \lambda, \mu \geqslant 0 \text{ si } \sigma_1 > \sigma_2 = \sigma_3 \,.$$
(1.10)

2. *COMPORTEMENT PLASTIQUE DES SOLS*

La définition d'un comportement plastique pour les sols présente, si l'on y regarde de près, de nombreuses difficultés et il pourrait sembler qu'il s'agisse là d'une schématisation abusive. Pourtant le succès des calculs d'équilibres limites dans la pratique de l'ingénieur, seuls disponibles de façon systématique lorsque l'on a à dégrossir un problème nouveau prouvent l'utilité de l'emploi raisonnable de ce modèle de comportement.

Il ne faut d'ailleurs pas perdre de vue que des difficultés analogues se présentent pour toute autre tentative de schématisation, simple pour être utilisable, du comportement des sols. L'essentiel est que le modèle de comportement adopté, soit adapté au problème à traiter.

2.1. Critère de plasticité

La figure 1, extraite d'un travail de Radenkovic, représentant des courbes de charge en coordonnées sans dimensions pour les sols et pour le métal, met clairement en évidence la différence de comportement entre les deux matériaux: pour les sols, les déformations irréversibles apparaissent pratiquement dès le début et il y a lieu, si on veut les suivre, d'adopter un modèle de comportement plastique avec écrouissage. Un palier est atteint pour des déformations importantes; c'est en fait à cette valeur, prise comme seuil, que l'on s'intéresse en général: le critère de plasticité utilisé correspond donc ainsi à un écoulement plastique bien établi, avec toutes les difficultés que cela comporte pour sa définition.

Le critère de plasticité adopté le plus souvent est le critère de Coulomb, sous forme courbe intrinsèque pour lequel:

$$f(\underline{\underline{\sigma}}) = \frac{\sigma_1 - \sigma_3}{2} + \frac{\sigma_1 + \sigma_3}{2} \sin\varphi - C\cos\varphi \qquad (2.1)$$

$$\sigma_1 \geqslant \sigma_2 \geqslant \sigma_3 , \qquad (> 0 \text{ si tractions}) .$$

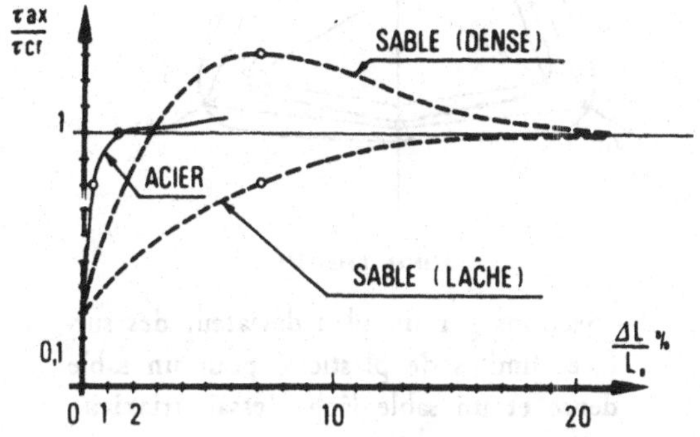

Fig. 1 – Courbes de charge pour acier, sable dense et sable lâche (densité plus élevée de 10% pour le sable dense)

D'autres formes de critères, apparentées à celle-ci, ont aussi été proposées rendant plus commodes certains calculs –critère de Drucker-Prager par exemple, qui fait intervenir la contrainte principale intermédiaire σ_2. Les résultats expérimentaux tels que ceux rapportés par Lade (1973) semblent indiquer que la forme (2.1), sans être exacte, traduit bien certaines propriétés des sols: rôle moindre de la contrainte principale intermédiaire.

Pour les argiles saturées non drainées le critère adopté est celui de Tresca, c'est à dire (2.1) dans lequel $\varphi = 0$.

Les travaux récents de Mandel sur le comportement plastique faisant intervenir la notion de trièdre directeur, devraient trouver dans les sols un vaste champ d'applications. Nous renverrons le lecteur au cours professé en 1971 au C.I.S.M. pour tous détails sur cette théorie qui n'a pour l'instant malheureusement reçu aucune application pratique.

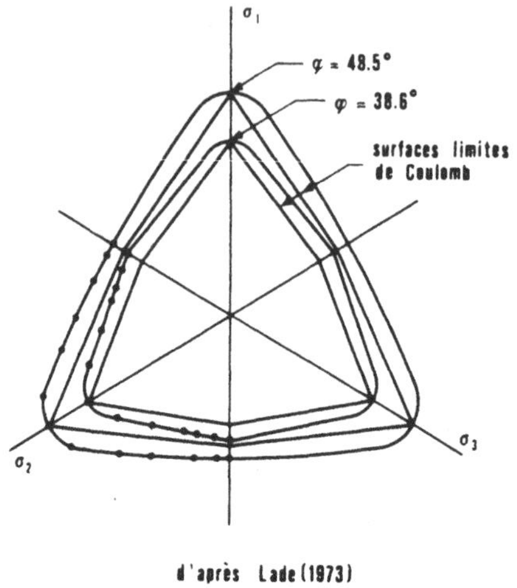

Fig. 2 – Sections par un plan déviateur des surfaces limites de plasticité pour un sable dense et un sable lâche (essais triaxiaux vrais).

2.2. Règle d'écoulement

Le problème de la régle d'écoulement plastique des sols est de loin le plus délicat.

Mettons tous de suite à part le cas des argiles saturées non drainées, pour lesquelles l'écoulement plastique se fait sans variation de volume et suivant le principe du travail maximal.

Il parait acquis par contre que pour les sols $\varphi \neq 0$, le principe du travail maximal n'est pas valable: il conduirait à une variation de volume–dilatance bien supérieure à ce qui est observé expérimentalement; lorsque la densité critique est atteinte la déformation plastique s'effectue pratiquement sans variation de volume.

Suivant Radenkovic nous classerons les approches du problème de la loi de comportement plastique des sols en deux groupes:

• approches "d'esprit cinématique", dans lesquelles on tente de construire la loi de comportement à partir de considérations sur la cinématique du milieu granulaire: le concept directeur est que la déformation s'effectue par glissement –en fait cisaillement– le long de facettes sur lesquelles la limite de frottement est atteinte. Les travaux de Mandel, Geniev, Spencer, de Josselin de Jong, ressortissent à cette catégorie. Il convient d'insister, comme Radenkovic, sur le fait que cette approche ne fournit pas une loi de comportement au sens classique puisqu'elle fait intervenir non seulement la vitesse de déformation $\underline{\underline{d}}$ mais aussi la vitesse de rotation $\underline{\underline{\omega}}$. Ceci explique les diverses difficultés rencontrées dans cette approche, liées au problème de la rotation (Mandel, Mandl et Luque, de Josselin de Jong).

Ici encore les notions de milieu polaire, ou milieu à directeurs introduit par Mandel, devraient dans l'avenir conduire à des résultats intéressants.

approches basées sur la notion de potentiel plastique: dans celles-ci, $\underline{\underline{d}}^p$ est défini comme dérivant d'un potentiel plastique g, distinct de f si le matériau n'obéit pas au principe du travail maximal. Cette idée, qui sous sa forme générale est due à Radenkovic (1962), avait été présentée par Jenike et Shield (1959) dans le cas d'un sol de Coulomb en utilisant le potentiel de von Mises d'où:

$$\underline{\underline{d}}^p = \lambda \underline{\underline{s}}, \qquad \lambda \geqslant 0 \tag{2.2}$$

soit dans le cas plan:

$$d_{xx} / \cos 2\theta = - d_{yy} / \cos 2\theta = d_{xy} / \sin 2\theta = \lambda \geqslant 0 \tag{2.3}$$

où $\theta = (0x, \sigma_1)$.

De même on a associé le critère de Coulomb à un potentiel de Tresca, en imposant de plus comme conséquence de l'isotropie du matériau, la coincidence des directions principales de $\underline{\underline{\sigma}}$ et de $\underline{\underline{d}}$.

Plus généralement Bent Hansen (1958), Radenkovic (1961), ont associé un critère de plasticité de Coulomb d'angle φ et un potentiel plastique, de Coulomb également, d'angle ν $0 \leqslant \nu \leqslant \varphi$, en imposant de plus la coincidence des directions principales due à l'isotropie, d'où:

$$d_1^p = \lambda(1 + \sin\nu), \; d_2^p = 0, \; d_3^p = -\lambda(1 - \sin\nu), \; \lambda \geqslant 0 \tag{2.4}$$

en régime de face $\sigma_1 > \sigma_2 > \sigma_3$

$$d_1^p = (\lambda + \mu)(1 + \sin\nu), \; d_2^p = -\mu(1 - \sin\nu), \; d_3^p = -\lambda(1 - \sin\nu) \tag{2.5}$$

$$\lambda, \mu \geqslant 0$$

en régime d'arête $\sigma_1 > \sigma_2 = \sigma_3$

Si $\nu = \varphi$, on retrouve le matériau de Coulomb standard et si $\nu = 0$ la déformation plastique s'effectue sans variation de volume, selon la règle de Tresca.

Une telle représentation paraît en accord avec certains résultats expérimentaux obtenus par l'Ecole de Cambridge.

Il est à signaler que la question de la coincidence des directions principales de $\underline{\underline{d}}$ et $\underline{\underline{\sigma}}$ comme conséquence de l'isotropie est fortement controversée: ainsi de Josselin de Jong admet une déviation entre les deux systèmes de directions principales caractérisée par un angle Φ.

Enfin, on doit aussi remarquer que dans (2.4 , 2.5), les conditions $\lambda \geqslant 0$, $\mu \geqslant 0$, sont plus restrictives que la condition thermodynamique de non négativité de la puissance dissipée (il y a équivalence pour $\nu = 0$).

On voit ainsi que les approches basées sur la notion de potentiel plastique font une très nette distinction entre les notions de

critère de plasticité (quand?)

et de règle d'écoulement (comment?) ;

tandis que les approches "d'esprit cinématique" mêlent intimement les deux notions, ce qui ne contribue apparemment pas à clarifier les choses. Aussi il semble que les approches du deuxième type soient pour l'instant les plus fructueuses.

Du point de vue expérimental, on doit mentionner les longues recherches effectuées par l'Ecole de Cambridge (Roscoe et al, 1967), par Goldscheider et Gudehus (1973), avec de vrais appareils triaxiaux. Ainsi le résultat remarquable suivant a été rapporté (figure 3): normalité de la projection de $\underline{\underline{d}}$ dans le plan déviateur à la section du critère de plasticité apparenté au critère de Coulomb comme on l'a dit au § 2.1; la déformation de volume pourrait être représentée par

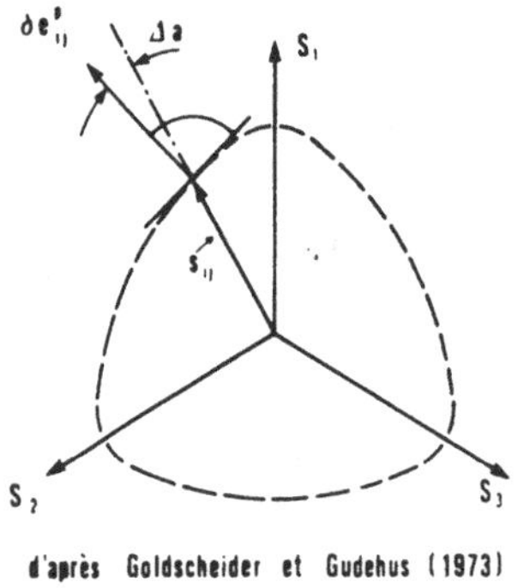

d'après Goldscheider et Gudehus (1973)

Fig. 3 – Section de la surface limite par le plan déviateur, et normalité dans ce plan.

une fonction de dilatance dépendant de la densité, (Gudehus, 1973).

Il convient aussi de signaler les travaux de Drescher et de Josselin de Jong pour étudier la cinématique des milieux granulaires sur modèles photoélastiques.

BIBLIOGRAPHIE

Pour une bibliographie complémentaire, voir chapitre I de l'ouvrage "Théorie de la Plasticité pour les Applications à la Mécanique des Sols".

A. DRESCHER et G. de JOSSELIN de JONG (1971), Photoelastic verification of a mechanical model for flow of granular media. J. Mech. Phys. Solids, vol. 20, p. 337-351.

G. GUDEHUS (1973), Elastoplastische Stoffgleichungen für trockenen Sand. Ingenieur--Archiv, 42, p. 151-169.

G.A. GENIEV (1958), Problèmes de la dynamique du milieu granulaire (en russe). Gostekhizd.

M. GOLDSCHEIDER & G. GUDEHUS (1973), Some sectionally proportional rectilinear extension tests on dry sand. Proc. Symp. on the role of Plasticity in Soil Mechanics, Cambridge (G.B.), 13-15 sept. 1973, p. 56-66.

A.W. JENIKE & R.T. SHIELD (1959), On the plastic flow of Coulomb solids beyond original failure. J. Appl. Mech. Trans. ASME, vol. 27, p. 599-602.

G. de JOSSELIN de JONG (1964), Lower Bound Collapse Theorem and Lack of Normality of Strain Rate to Yield Surface for Soils. Proc. IUTAM Symp. on Rheology and Soil Mech., Grenoble, p. 69-75.

P.V. LADE (1973), Discussion, session II, Proc. Symp. on the role of Plasticity in Soil Mechanics, Cambridge (G.B.), 13-15 sept. 1973, p. 129-135.

J. MANDEL (1947), Sur les lignes de glissement et le calcul des déplacements dans la déformation plastique. C.R. Ac. Sc., Paris, t. 225, p. 1272-1273.

J. MANDEL (1966), Sur les équations d'écoulement des sols idéaux en déformation plane et le concept du double glissement. J. Mech. Phys. Solids, 14, p. 303-308.

J. MANDEL (1971), Plasticité Classique et Viscoplasticité. Cours au C.I.S.M. 1971, Springer-Verlag.

G. MANDL & R.F. LUQUE (1970), Fully developped plastic shear flow of granular materials. Géotechnique, 20, p. 277-307.

D. RADENKOVIC (1961), Théorèmes limites pour un matériau de Coulomb à dilatation non standardisée. C.R. Ac. Sc., Paris, 252, p. 4103-4104.

D. RADENKOVIC (1962), Théorie des charges limites. in Séminaire de Plasticité, (ed. J. Mandel), p. 129-142.

D. RADENKOVIC (1972), Equilibre limite des milieux granulaires. Modèles de comportement rigide-plastique. in Plasticité et Viscoplasticité 1972, ed. D. Radenkovic et J. Salençon, Ediscience, Paris 1974, p. 379-394.

K.H. ROSCOE et al. (1967), Principal axes observed during simple shear of a sand. Proc. Geot. Conf., Oslo, p. 231-238.

PROBLEMES D'ELASTOPLASTICITE EN MECANIQUE DES SOLS

1. PROBLEMES D'ELASTOPLASTICITE

Pour un milieu élastoplastique, la loi de comportement s'obtient en écrivant que:

$$\underline{\underline{d}} = \underline{\underline{d}}^e + \underline{\underline{d}}^p \qquad (1.1)$$

somme des parties élastique et plastique, cette dernière étant définie par la règle d'écoulement plastique.

Par suite du caractère incrémental de cette loi, on doit pour déterminer l'état de contraintes et de déformations dans un état de charge donné, suivre (et donc connaître) tout le trajet de charge qui aboutit à cet état. La résolution de tels problèmes est donc longue; il est rare que l'on puisse procéder par voie analytique et l'on a en règle générale recours à des méthodes numériques, à propos desquelles des problèmes de convergence du schéma numérique dans le temps peuvent se poser (Nguyen et Zarka, 1972).

Dans les dernières années de nombreux problèmes d'élastoplasticité ont ainsi été résolus numériquement, en général dans le cas de matériaux de Tresca ou de Mises à écrouissage positif ou même négatif; le matériau de Coulomb ayant aussi été abordé avec l'inévitable problème dû à la règle d'écoulement, certains auteurs prenant la règle de normalité, d'autres telle ou telle hypothèse présenté au chapitre précédent.

Tous les calculs numériques ont été menés dans l'hypothèse des petites déformations, sans tenir compte des changements de géométrie.

Les calculs d'élastoplasticité nécessitant comme nous l'avons dit la connaissance du trajet de charge –donc des contraintes initiales– l'utilisation de tels calculs en mécanique des sols n'est pas évidente, à la différence du cas du métal.

2. *CHARGE CRITIQUE – CHARGE LIMITE*

Nous allons à partir d'un exemple essayer de dégager les différents comportements possibles pour un système en matériau élastoplastique soumis à un processus de chargement.

2.1. Enveloppe sphérique soumise à une pression interne

Nous étudions d'abord le cas d'une enveloppe sphérique de rayons initiaux intérieur et extérieur a_o et b_o, soumise à une pression interne croissante p, la pression extérieure étant nulle. Les contraintes initiales sont nulles. L'enveloppe est constituée d'un matériau de Tresca standard, sans écrouissage.

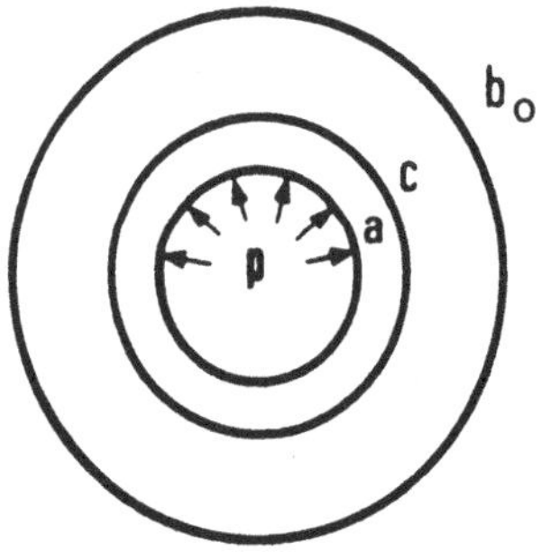

Fig. 1 – Enveloppe sphérique soumise à une pression interne

En supposant les déformations petites, on obtient après des calculs classiques la pression p en fonction du rayon c de la frontière entre zone élastique et zone plastique.

$$p = 4C\left[\frac{1}{3}\left(1 - \frac{c^3}{b_o^3}\right) + \ln\frac{c}{a_o}\right] \tag{2.1}$$

formule ne faisant pas intervenir E et ν, ce qui est une particularité due aux propriétés de la solution élastique de Lamé.

La limite d'élasticité du système étudié sous ce type de chargement, définie comme correspondant à l'apparition de la plasticité en au moins un point, correspond à $c = a_o$ et est:

$$p_o = \frac{4}{3}C\left(1 - \frac{a_o^3}{b_o^3}\right) . \tag{2.2}$$

La plastification totale de l'enveloppe correspond à $c = b_o$ et est atteinte pour:

$$p = p_\ell = 4C \ln \frac{b_o}{a_o}. \tag{2.3}$$

C'est la pression maximale possible car les déformations plastiques qui étaient jusque-là limitées par la couche élastique qui entourait la zone plastique ($a_o < r < c$) deviennent alors libres. La déformation se poursuivrait alors sous charge constante si l'on ne tenait pas compte des changements de géométrie.

p_ℓ est apelée *la pression limite* du système.

Remarquons que cette valeur est susceptible d'une autre définition. Les changements de géométrie étant négligés, l'étude est menée sur la géométrie initiale du système. Aussi, si l'on considère le même système constitué du même matériau plastique, *sans* élasticité, p_ℓ est la pression pour laquelle ce système, jusque-là indéformé, commence à se déformer, les déformations étant purement plastiques. On comprend alors pourquoi p_ℓ ne dépend pas (c'est une règle gènérale) des caractéristiques élastiques du système. p_ℓ est la pression d'apparition de l'écoulement plastique libre commençant.

Les résultats précédents sont valables si l'hypothèse des déformations petites, jusqu'à ce que p atteigne p_ℓ, est acceptable. Le calcul fait par Mandel (1966) montre qu'il en est bien ainsi, si:

$$\frac{C}{E}(1-\nu)\left(\frac{b_o}{a_o}\right)^3 << 1. \tag{2.4}$$

Dans le cas d'une enveloppe épaisse, l'étude doit être menée en tenant compte des changements de géométrie. La figure 2 représente l'évolution du rayon intérieur de l'enveloppe, a, en fonction de p.

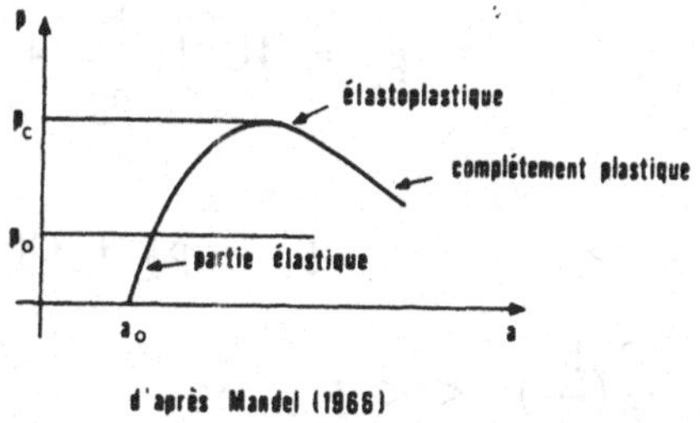

Fig. 2 – Enveloppe sphérique sous pression interne: évolution de la pression en fonction du rayon intérieur (grandes déformations).

Il existe un chargement critique p_c, au-delà duquel la déformation croît sous pression interne décroissante, (Instabilité). Prenant pour simplifier le cas du matériau incompressible, on a:

en posant

$$\kappa = [1 + \frac{12C}{E} (\frac{b_o}{a_o})^3]^{1/2}$$

$$(2.5) \qquad p_c = \frac{4C}{3}[1 + \ln \frac{E}{3C} - \frac{E}{6C}(\frac{a_o}{b_o})^3 (\kappa - 1) + \ell n \frac{\kappa - 1}{\kappa + 1}]$$

p_c est atteinte avant plastification complète de l'enveloppe sphérique, la plastification se poursuit ensuite sous pression décroissante jusqu'à devenir complète ou bien la déformation se localise. Il est clair que dans ce problème c'est p_c , pression critique, qui joue le rôle important de pression interne maximale possible.

Remarquons que dans ce cas, la pression p_ℓ donnée par (2.3), pression d'apparition de l'écoulement plastique libre dans le système rigide plastique associé, n'a aucune signification.

Ce n'est que pour une enveloppe sphérique mince,

$$\frac{C}{E} (\frac{b_o}{a_o})^3 < < 1$$

que les deux notions se rejoignent:

p_c vient se confondre avec p_ℓ

$$(2.6) \qquad p_c = 4C \ln \frac{b_o}{a_o} + O(\frac{C}{E} (\frac{b_o}{a_o})^3)$$

$$p_c = p_\ell + O(\frac{C}{E} (\frac{b_o}{a_o})^3)$$

pour $\frac{C}{E}(\frac{b_o}{a_o})^3 < < 1.$

2.2. Considérations générales

D'une façon générale, dans le cas d'un système en matériau élastique parfaitement plastique, soumis à un processus de chargement croissant dépendant d'un paramètre Q, on passe successivement par les phases suivantes:

- phase élastique justqu'à la limite d'élasticité du système

$$0 < Q < Q_o$$

- phase plastique

$$Q_o \leq Q$$

Pour celle-ci, on a vu sur l'exemple précédent qu'elle peut présenter des allures différentes suivant les caractéristiques géométriques et matérielles du système étudié. D'une façon générale, sans que nous puissions le démontrer, nous pensons que l'analyse élastoplastique du problème, sans négliger les changements de géométrie, conduit pour la variable de charge Q, et la variable de déplacement associée q à une évolution qui présente l'une des 3 formes suivantes:

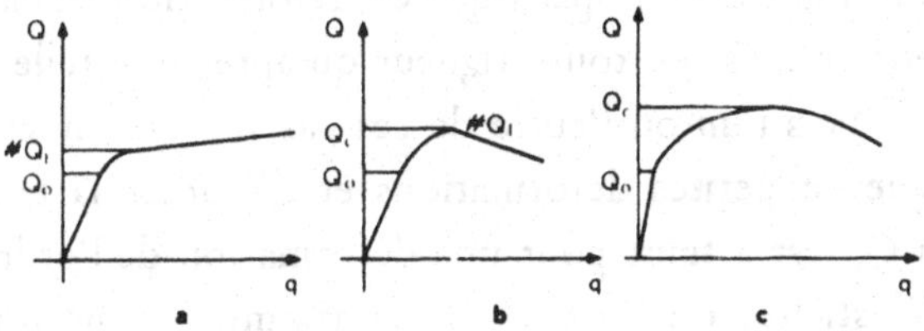

Fig. 3

– Dans la forme 3a, la partie élastoplastique présente d'abord un coude assez prononcé, $Q > Q_0$, pour aboutir ensuite à une montée de pente relativement faible due uniquement à "l'écrouissage géométrique du système". Le début de cette montée est très voisin de Q_1, chargement limite, calculé en négligeant les changements de géométrie, c'est à dire en particulier comme chargement correspondant à l'apparition de l'écoulement plastique libre dans le système rigide plastique associé.

Q_1 a donc une signification physique: il correspond pratiquement à l'apparition des grandes déformations dans le système. Jusqu'à Q_1 l'étude peut être menée en petites déformations.

– Dans la forme 3b, la partie élastoplastique passe par un maximum pour $Q = Q_c$, atteint pour une déformation qui reste de l'ordre de grandeur de celle à la limite d'élasticité; l'écoulement plastique libre (c'est à dire non contenu) apparaît pour une valeur de Q, inférieure à Q_c mais très voisine.

Dans ce cas Q_c est en fait très voisine de Q_1 défini comme correspondant à l'écoulement plastique libre dans le système rigide plastique associé.

Q_1 a donc encore une signification physique: c'est pratiquement la charge maximale que peut supporter le système. L'étude peut être menée en petites déformations pratiquement jusqu'à Q_1.

– Dans la forme 3c, la partie élastoplastique présente un maximum pour $Q = Q_c$ atteint pour une déformation nettement supérieure à celle à la limite d'élasticité.

Q_1 n'a alors plus aucune signification physique. L'analyse élastoplastique en grande déformation est indispensable.

Il résulte de cela que si l'on s'intéresse à la charge ultime de la structure, on pourra dans les deux premiers cas se réfèrer à Q_1, qui peut être obtenue à partir du schéma de comportement rigide plastique, tandis que dans le troisième cas il faudra déterminer Q_c par une analyse élastoplastique en déformations finies.

Malheureusement ce n'est en toute rigueur qu'après une telle analyse que l'on peut savoir si l'on est dans l'un ou l'autre de ces cas. En fait, il suffira de faire une analyse élastoplastique en petites déformations et d'étudier la courbe (Q, q) ainsi obtenue: si, ou bien Q_1 est atteint pour une déformation de l'ordre de grandeur de celle à la limite d'élasticité, ou bien Q_1 apparaissant comme une valeur asymptotique, la courbe (Q, q) "colle" vite à son asymptote, on se trouve dans le ler cas.

Bien souvent, d'ailleurs, on se bornera à faire appel à l'intuition, ce qui peut présenter quelques dangers.

3. *CAS DE LA MECANIQUE DES SOLS – UN PROBLEME D'APPLICATION COURANTE*

3.1. Problèmes d'élastoplasticité en Mécanique des Sols

La Mécanique des Sols a fait, jusqu'ici, peu appel aux problèmes d'élastoplasticité. Nous pensons que l'on peut trouver à cela plusieurs raisons, parmi lesquelles les suivantes:

la nécessité de connaître les contraintes initiales et le trajet de charge correspond mal à la réalité des problèmes,

la schématisation élastoplastique du comportement n'est pas suffisamment valable pour justifier des calculs coûteux, et on préfère adopter la schématisation rigide plastique qui est beaucoup plus efficace et calculer les charges limites, qui, lorsqu'elles ont un sens, correspondent mieux aux résultats cherchés.

Pour certains problèmes néanmoins, l'analyse élastoplastique peut se révéler utile: soit que le trajet de charge soit suffisamment bien connu et que l'on s'intéresse à la réponse du système bien avant la charge ultime, soit que s'intéressant à la charge ultime on se trouve dans un cas où la schématisation rigide plastique est inapplicable.

Nous donnons au § suivant l'exemple d'un type de problèmes dont la résolution élastoplastique dans l'hypothèse des grandes déformations est très utile pour de nombreuses questions en Mécanique des Sols.

3.2. Cavité en milieu élastoplastique infini

Le problème étudié est celui d'une cavité sphérique ou cylindrique soumise à une pression interne dans un milieu élastoplastique infini où règne une contrainte initiale (pression à l'infini P). L'analyse de ce problème, effectuée en grandes déformation par Mandel (1966) et par l'auteur (1966, 1969) a été menée dans le cas du matériau de Tresca standard, et du matériau de Coulomb standard ou non standard.

Ce problème schématique se rattache à de nombreuses applications en Mécanique des Sols: tenue de souterrains à grande profondeur, cavités souterraines pour stockage à grande profondeur, appareils pressiométriques, etc. . .

Sans entrer dans les détails des calculs longs et pénibles, nous donnons à la figure 4, dans le cas de la cavité cylindrique, la représentation de la relation entre la pression interne p et le rayon actuel de la cavité (Habib, 1973, d'après les travaux de l'auteur). Les résultats sont entièrement analogues dans le cas d'une cavité sphérique.

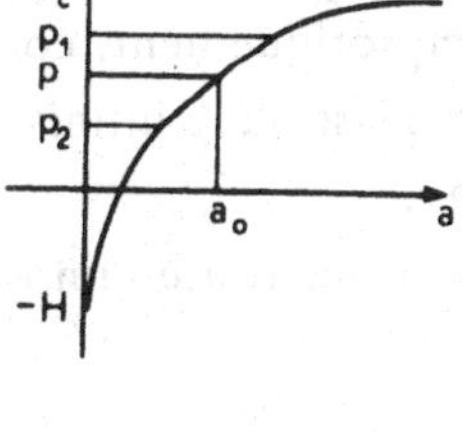

Fig. 4 – Variation du rayon d'une cavité en fonction de la pression interne (élastoplasticité en déformation finies)

Le tableau ci-dessous tiré de Habib (1973) résume les diverses phases du comportement de la cavité.

p croissant à partir de P on a d'abord une phase élastique; puis pour $p > p_1$ la phase élastoplastique dans laquelle les déformations plastiques se produisent à partir de la cavité. cette phase est limitée pour la pression par la valeur p_c

p décroissant à partir de P on a d'abord une phase élastique; puis pour $p < p_2$, phase élastoplastique où les déformations se produisent à partir de la cavité; suivant les cas p_2 est positive ou négative (traction); pour $p = -H = -C \cot g \varphi$ la cavité est complètement écrasée.

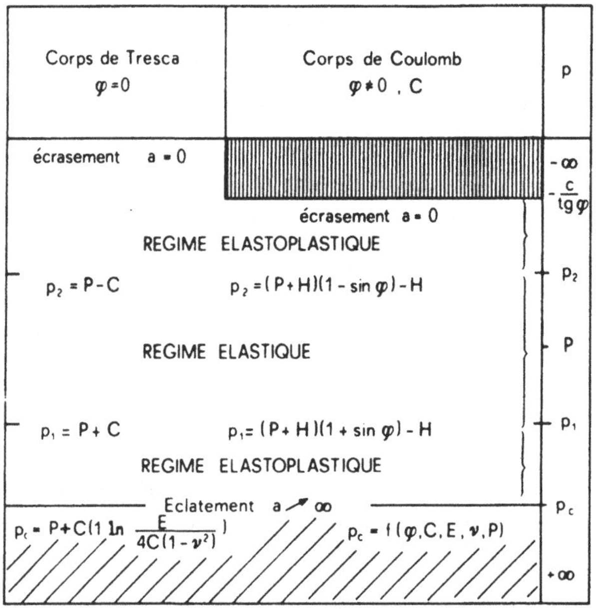

On remarque qu'il existe pour p une valeur maximale, atteinte asymptotiquement. Cette valeur est notée p_c. Il s'agit bien en effet de l'homologue de la pression critique introduite au § 2.1., et non d'une pression limite au sens de Q_1 (§ 2.2): on a affaire à une enveloppe cylindrique infiniment épaisse ($b_o = \infty$), pour laquelle p_ℓ donnée par une formule analogue à (2.3) est infinie, et p_c est atteint asymptotiquement, correspondant à a et c (rayon de la frontière entre zones élastique et plastique) infinis. La formule donnant p est analogue à (2.5) où l'on fait $b_o = \infty$.

On voit donc toute l'importance pratique de mener un tel calcul élastoplastique en déformations finies. De plus des formules assez compliquées mais parfaitement exploitables permettent aussi de connaitre en fonction de p le rayon de la zone plastique, ce qui aidera à évaluer la tenue de l'ouvrage étudié.

BIBLIOGRAPHIE

F. BAGUELIN et al. (1972), Expansion of cylindrical probes in cohesive soils. J. Soil Mech. and Found. Div., A.S.C.E., vol. 98, n. SMI, nov. 1972, p. 1129-1142.

P. HABIB (1973), Précis de Géotechnique, Dunod, Paris, 1973.

J. MANDEL (1966), Mécanique des Milieux Continus. Tome II, Gauthier- Villars, Paris, 1966.

Q.S. NGUYEN et J. ZARKA (1972), Quelques méthodes de résolution numérique en plasticité classique et en viscoplasticité. in Plasticité et Viscoplasticité 1972, p. 327-357, ed. par D. Radenkovic et J. Salençon, Ediscience, Paris, 1974.

D. RADENKOVIC et J. SALENÇON (1971), Equilibre limite et rupture en mécanique des sols. in Le comportement des sols avant la rupture, Journées nationales du C.F.M.S. 1971, p. 296-302, n. Spécial Bulletin de Liaison des Laboratoires des Ponts et Chaussées.

J. SALENÇON (1966), Expansion quasi-statique d'une cavité . . . dans un milieu élasto-plastique. Ann. Pts et Ch., III, 1966, p. 175-187.

J. SALENÇON (1969), Contraction quasi-statique d'une cavité . . . Ann. Pts et Ch., IV, 1969, p. 231-236.

A.S. VESIC (1972), "Expansion of cavities in an infinite soil mass". J. Soil. Mech. and Found. Div., A.S.C.E., vol. 98, n. SM3, mars 1972, p. 265-290.

SCHEMA RIGIDE-PLASTIQUE
CHARGEMENTS LIMITES

1. *LE MATERIAU RIGIDE-PLASTIQUE*

Le schéma de comportement rigide-plastique est de loin le plus utilisé en Mécanique des Sols.

Par définition un matériau rigide-plastique est un matériau pour lequel il n'y a pas de déformation élastique, soit:

$$\underline{\underline{d}} = \underline{\underline{d}}^{p} \tag{1.1}$$

Nous avons introduit ce schéma de comportement, incidemment au chapitre précédent lorsque, considérant des sytèmes élasto parfaitement plastiques pour lesquels les changements de gèométrie étaient négligeables jusqu'à (ou presque jusqu'à) l'apparition de l'écoulement plastique libre, nous avons parlé de chage limite.

Il est important de remarquer que le schéma de comportement rigide-plastique va, en général, conduire à des indéterminations dans la résolution des problèmes; en effet, le matériau étant indéformable hors les zones plastiques, il sera le plus souvent impossible de déterminer les contraintes hors ce zones.

Il ya lieu de considérer que l'on a affaire à la limite d'un matériau élastoplastique lorsque l'on fait tendre les modules d'élasticité vers l'infini: le résultat dépend a priori évidemment du matériau élastique de départ et de la façon dont a passe à la limite.

Considérons un système en matériau élasto-parfaitement plastique soumis à un processus de chargement et faisons tendre ce système vers le système rigide parfaitement plastique associé en multipliant tous les modules d'élasticité par un facteur $\rho >$ 0 qui $\nearrow \infty$. Alors:
en général –mais il y a des exceptions telles que le problème de la cavité en milieu infini évoqué au chapitre précédent– pour ρ suffisamment grand, les changements de géométrie deviennent négligeables jusqu'à (ou presque) l'apparition de l'écoulement plastique libre et on démontre (cf. par exemple, l'ouvrage cité de l'auteur) que la charge d'apparition de l'écoulement plastique libre ne dépend alors

plus de ρ, et peut être en particulier déterminée en faisant ρ infini, c'est à dire sur le système rigide plastique associé: charge limite.

Il va de soi que le système rigide parfaitement plastique associé ne peut avoir de déformation non nulle qu'à partir du stade de l'écoulement plastique libre commençant ; on peut si on le désire poursuivre l'étude de l'écoulement plastique libre ensuite en tenant compte des changements de géométrie (écrouissage géométrique du système).

En Mécanique des Sols le schéma rigide-plastique est utilisé uniquement pour la détermination de chargements limites. Il pourrait sembler quelque peu paradoxal de se préoccuper d'écoulement plastique libre commençant alors que nous avons dit antérieurement que le critère de plasticité et la phase "parfaitement plastique" du comportement des sols correspondaient à la déformation plastique établie du matériau. En fait, nous verrons que l'on dispose pour la détermination des chargements limites, de moyens d'approche puissants et l'expérience montre que les résultats ainsi obtenus sont précieux pour le mécanicien des sols.

2. *PROCESSUS DE CHARGEMENT A N PARAMETRES*

Comme nous l'avons dit, le schéma de comportement rigide-plastique n'a d'utilisation que pour l'étude des problèmes d'écoulement plastique libre commençant: on va chercher à déterminer la valeur de la charge appliquée au système pour laquelle il y a écoulement plastique libre. Ceci implique des conséquences essentielles sur la façon dont les donnée du problème doivent être exprimées.

En effet il est clair, puisque l'on cherche à avoir l'écoulement plastique libre, que le données doivent être telles que cet écoulement soit possible, et la résolution du problème aura pour but la détermination de la charge:
ainsi, les donées en forces (au contour, et de masse) du problème doivent présenter un arbitraire suffisant pour que l'écoulement plastique libre soit possible.

La formulation correcte des données des problèmes pour les systèmes rigides-plastiques est fondamentale. Elle repose sur la notion de paramètres de chargement.

Jusqu'ici, partant du système élastoplastique nous nous sommes attachés à suivre un trajet de charge prescrit, sur lequel nous aboutissons au chargement limite pour ce trajet; le plus souvent d'ailleurs (cf. l'exemple du chap. 2) il s'agissait de trajets de charge dans lesquels les forces appliquées croissent proportionellement entre elles. Cela n'est pas suffisamment général, par contre l'ensemble des problèmes plastiques est correctement représenté en introduisant les processus de chargement

dépendant de n paramètres (n fini):
par exemple, une fondation rigide lisse agissant à la surface d'un sol homogène de Tresca non pesant représente un processus de chargement dépendant de 2 paramètres, c'est à dire que tous les trajets de charge possibles pour ce système peuvent être représentés dans un espace à 2 dimensions: l'effort normal et le moment de renversement de la fondation.

Fig. 1 – Fondation lisse agissant à la surface d'un sol pesant

On trouvera dans l'ouvrage cité de l'auteur, une présentation formelle détaillée de cette notion de paramètres de chargement. Le résultat fondamental est le suivant qui a trait à l'expression du principe des puissances virtuelles:

si on a affaire à un système qui est soumis à un processus de chargement à n paramètres, alors pour tout champ de contrainte $\underline{\underline{\sigma}}$ statiquement admissible pour le processus étudié,

pour tout champ de vitesse $\underline{v}^*$, vitesse de déformation $\underline{\underline{d}}^*$, cinématique admissible pour le processus étudié, on a:

$$\int_\Omega \operatorname{tr}(\underline{\underline{\sigma}} \cdot \underline{\underline{d}}^*)\, d\Omega = \sum_{j=1}^{n} Q_j(\underline{\underline{\sigma}}) \cdot \dot{q}_j^*(\underline{v}^*) = \underline{Q}(\underline{\underline{\sigma}}) \cdot \underline{\dot{q}}^*(\underline{v}^*) \qquad (2.1)$$

où les Q_j sont les paramètres de chargement et les $\dot{q}_j$ les paramètres de déformation du système, associés.

Les correspondances $\underline{\underline{\sigma}} \to \underline{Q}(\underline{\underline{\sigma}})$ et $\underline{v}^* \to \underline{\dot{q}}(\underline{v}^*)$ sont linéaires.

Tout ce qui a été dit auparavant à propos d'un trajet de charge est valable pour tout trajet de charge du processus à n paramètres: en particulier en ce qui concerne la notion de chargement limite. On a ainsi dans l'espace des paramètres de chargement un ensemble de chargements limites dont le lieu est appelé "frontière d'écoulement du système" –yield locus for the system–, dénomination qui est justifiée par la théorie des charges limites dans le cas du matériau standard à fonction de charge convexe.

A titre d'exemple revenant à la figure 1, dans l'espace (M,N), on a un ensemble de points limites, qui constitue la frontière d'écoulement pour ce problème.

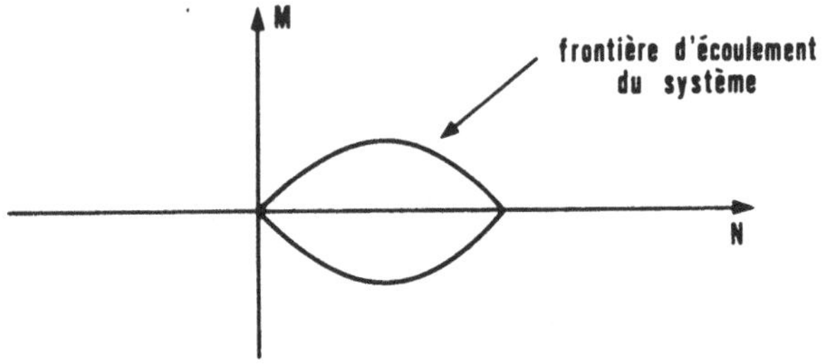

Fig. 2 – frontière d'écoulement du système de la fig. 1

THEORIE DES CHARGES LIMITES POUR LES SYSTEMES STANDARDS

1. SYSTEME STANDARD

Les systèmes mécaniques auxquels on a affaire sont en général constitués de plusieurs solides en contact le long d'interfaces.

Plutôt que de considérer chaque solide isolément avec des données aux limites sur les interfaces de contact définies par les conditions de frottement correspondantes, nous pensons plus commode de considérer le système global formé de tous les solides, avec des données aux limites du type indiqué au chapitre précédent, c'est à dire définies par un certain nombre de paramètres de chargement. Dans cette vision, les interfaces de contact entre les solides constitutifs du système, sont des éléments du système (analogues à des éléments de matière), dont la condition de frottement est la loi de comportement plastique.

On est donc amené, de la même façon que pour les matériaux, à distinguer les interfaces standards et non standards. On trouvera dans l'ouvrage cité de l'auteur, une étude détaillée des interfaces, dont nous rappellerons ici quelques résultats principaux.

L'interface à contact *lisse,* avec décollement possible est standard.

L'interface parfaitement rugueux, c'est à dire à adhérence parfaite est standard sans ambiguité s'il se trouve entre deux matériaux de Tresca (ou Misès) standards. De même pour les interfaces partiellement rugueux.

L'interface avec condition de frottement sec de Coulomb n'est pas standard (il y a bien convexité mais pas normalité).

La définition d'un système standard es alors évidente: un système est standard si la loi de comportement de tous ses éléments est standard.

2. BUT DE L'ANALYSE LIMITE

Le but de l'analyse limite est la détermination des chargements limites. Travaillant directement sur le matériau rigide-plastique, on laisse de côté le trajet de charge et la détermination pas à pas, pour procéder directement par des méthodes de type variationnel.

Les théorèmes de l'analyse limite sont exposés dans de nombreux ouvrages. Nous nous bornerons à rappeler les notions fondamentales et les énoncés, renvoyant le lecteur par exemple à l'ouvrage cité de l'auteur, pour détails fins et démonstrations.

APPROCHE STATIQUE

3.1. L'idée de base

Les chargements limites cherchés, sont les chargements pour lesquels il y a déformation du système rigide-plastique, c'est à dire auxquels il correspond une solution d'équilibre limite (champ de contrainte $\underline{\underline{\sigma}}$, champ de vitesse de déformation $\underline{\underline{d}} \neq 0$).

L'idée fondamentale de l'approche statique est de déterminer les chargements limites par la seule considération des champs de contraintes en mettant en évidence une propriété caractérisant les champs de contraintes des solutions d'équilibre limite parmi les champs admissibles, et les chargements correspondants.

3.2. Champs de contraintes licites ou S.P.A.

Un champ $\underline{\underline{\sigma}}$ est licite ou Statiquement et Plastiquement Admissible si:

$$\left\{ \begin{array}{ll} \underline{\underline{\sigma}} = & \text{champ statiquement admissible pour le processus} \\ \text{et} & f(\underline{\underline{\sigma}}) \leqslant 0 \text{ partout.} \end{array} \right. \tag{3.1}$$

3.3. Chargements licites

Un chargement $\underline{Q}$ du processus est licite s'il est équilibré par au moins un champ $\underline{\underline{\sigma}}$ licite.

3.4. Il résulte de la *convexité* du critère pour tous élément du système, que l'ensemble des chargements licites, K, est convexe dans l'espace R^n des paramètres de chargement.

3.5. Il résulte de plus de la règle de *normalité* que les chargements limites sont sur la frontière de K et réciproquement.

3.6. Conséquences

La frontière de K est la frontière d'écoulement du système, ce qui justifie d'ailleurs cette dénomination (il s'agit bien d'une surface dans l'espace R^n).

Dans cette approche les chargements limites apparaissent indépendamment de toute considération de trajet de charge et de matériau élastoplastique. Ceci a deux conséquences importantes:

- le matériau rigide-plastique est *unique* quels que soient le matériau élastoplastique de départ et la façon dont on effectue le passage à la limite, lorsque l'on envisage les chargements limites pour le matériau standard.
- les chargements limites sont indépendants du trajet suivi pour les atteindre.

3.7. Théorème statique – Méthode statique

Il découle des résultats précédents que:

un chargement licite (c'est à dire équilibré par au moins un champ de contraintes licite) est un chargement stable. (Evidemment tout chargement stable est un chargement licite).

Par application de la propriété de convexité de K, on est conduit à une approximation par l'intérieur de la frontière d'écoulement, (fig. 1).

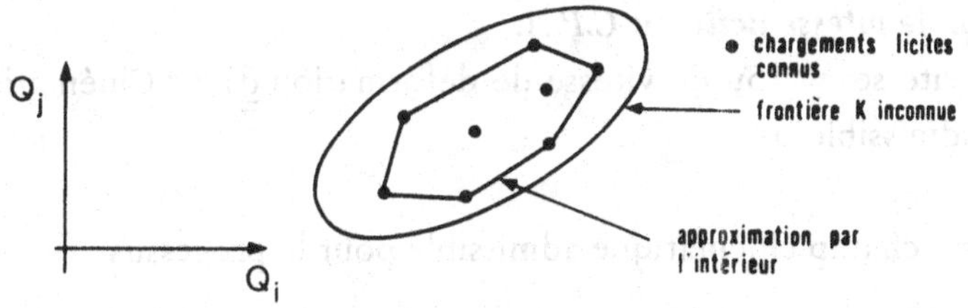

Fig. 1 – Méthode statique

Dans le cas d'un paramètre de chargement unique, K est un segment dont les extrémités sont les *deux* chargements limites.

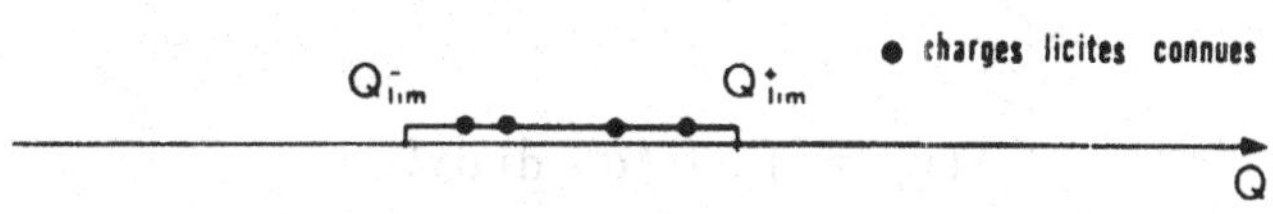

Fig. 2

et on a:

$$(3.2) \qquad Q^{+}_{\text{lim}} = \text{Max}\ \{\ Q(\underline{\underline{\sigma}})\ ,\ \ \underline{\underline{\sigma}}\ \text{licite}\ \}\ .$$

3.8. Remarque finale

Il est important de garder en mémoire que malgré leur aspect intuitif correspondant à un mode de raisonnement utilisé naturellement depuis des siècles, le théorème statique et la méthode statique reposent sur les *deux* hypothèses de convexité et de normalité.

4. APPROCHE CINEMATIQUE

4.1. L'idée de base

Déterminer la frontière d'écoulement du système par la seule considération des champs de vitesse admissibles en mettant en évidence une propriété caractérisant les chargements limites par rapport à ces champs.

4.2. Champ de vitesse licite ou C.P.A.

Un champ vitesse $\underline{v}$ (ou de vitesse de déformation $\underline{\underline{d}}$) est Cinématiquement et Plastiquement Admissible si:

$$(4.1) \qquad \begin{cases} \underline{v} = \text{champ cinématique admissible pour le processus} \\ \qquad \text{et} \\ \exists\ \underline{\underline{\sigma}}\ \text{P.A. tel que}\ \underline{\underline{d}} \in \lambda\, \partial f(\underline{\underline{\sigma}}),\ \lambda \geqslant 0\ \text{en tout point} \end{cases}$$

($\underline{\underline{\sigma}}$ n'est pas nécessairement S.A.)

Autrement dit, le champ $\underline{v}$ vérifie les conditions nécessaires pour correspondre à une déformation plastique du système dans le cadre du processus étudié.

On démontre que, quel que soit le champ $\underline{\underline{\sigma}}$ associé à $\underline{v}$ dans (4.1) l'intégrale:

$$(4.2) \qquad P(\underline{v}) = \int_{\Omega} \text{tr}(\underline{\underline{\sigma}} \cdot \underline{\underline{d}})\, d\Omega$$

est une fonction univoque de $\underline{v}$: c'est la puissance dissipée associée au champ $\underline{v}$.

Le champ $\underline{v}$ peut admettre des discontinuités le long de lignes ou de surfaces isolées, sous réserve que la discontinuité vérifie une condition pour être P.A. (par ex.: discontinuité tangente à la ligne, dans le cas de Tresca ou de Mises); la formule (4.2), que l'on peut comprendre au sens des distributions, comporte alors un terme de volume et une terme de surface; ainsi dans le cas du matériau de Mises:

$$P(\underline{v}) = \int_{\Omega} k\sqrt{2 \operatorname{tr} \underline{\underline{d}}^2} \; d\Omega + \int_s k \, | [\underline{v}] | \, ds \, . \qquad (4.3)$$

4.3. Théorème cinématique – Méthode cinématique

On peut alors énoncer le théorème cinmatique de la théorie des charges limites dont la démonstration repose sur les hypothèses de convexité et normalité:

Tout chargement dont la puissance dans un champ de vitesses licite est supérieure à la puissance dissipée dans ce champ de vitesses, est un chargement instable.

On en déduit la méthode cinématique de détermination de la frontière d'écoulement, qui fournit une approximation par l'extérieur de cette frontière:

pour chaque $\underline{v}$ le demi espace défini dans l'espace des chargements par:

$$\underline{Q} \cdot \dot{\underline{q}}(\underline{v}) - P(\underline{v}) \geqslant 0 \qquad (4.4)$$

est extérieur ou tangent à K.

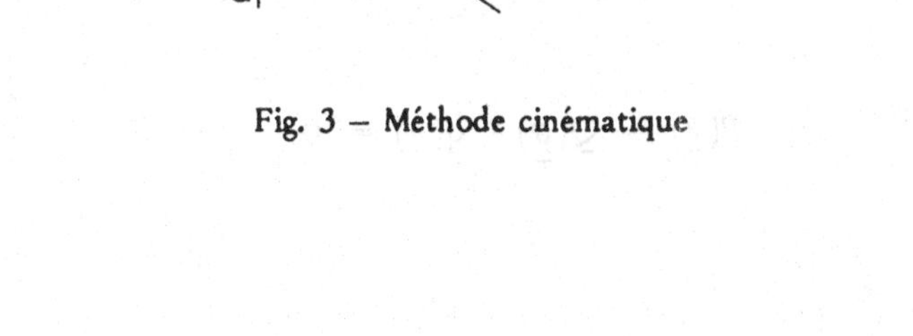

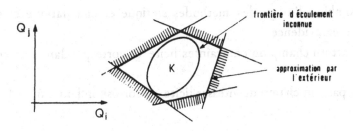

Fig. 3 – Méthode cinématique

Dans le cas d'un paramètre de chargement unique:

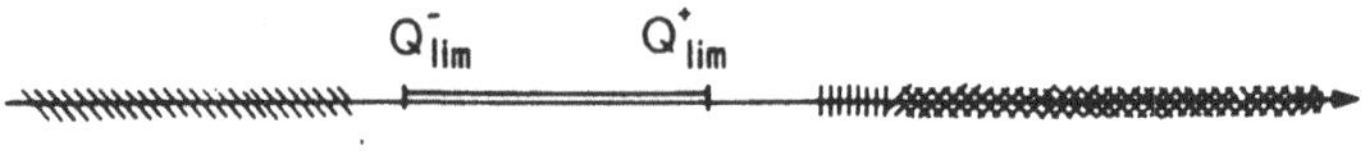

Fig. 4

et on a:

(4.5) $$Q^{+}_{\lim} = \mathrm{Min}\ \{\ P(\underline{v})\ |\ \underline{v}\ \text{licite},\ \dot{\underline{q}}(\underline{v}) = +1\}\ .$$

4.4. Remarque finale

Comme le théorème statique, le théorème cinématique énoncé ci-dessus a un caractère intuitif qui correspond à une mode de raisonnement couramment utilisé. Il est donc essentiel de garder en mémoire que ce théorème repose sur les hypothèses de convexité et normalité.

5. THEOREME D'ASSOCIATION – THEOREME D'UNICITE

5.1. Théorème d'association – Méthode combinée

Le théorème suivant est une conséquence des hypothèses de convexité et normalité:

si l'on combine l'usage des méthodes statique et cinématique et si, par ces moyens, on met en évidence:

· d'une part un champ de contraintes licite $\underline{\underline{\sigma}}$ correspondant à un chargement $\underline{Q}(\underline{\underline{\sigma}})$

· d'autre part un champ de vitesses licite $\underline{v}$ correspondant à $\dot{\underline{q}}(\underline{v})$

tels que:

$$P(\underline{v}) - \underline{Q}(\underline{\underline{\sigma}}) \cdot \dot{\underline{q}}(\underline{v}) = 0$$

alors

- $\underline{Q}(\underline{\underline{\sigma}})$ est chargement limite
- $\underline{\dot{q}}(\underline{v})$ est la vitesse de déformation du système associée
- les champs $\underline{\underline{\sigma}}$ et $\underline{v}$ sont associé et constituent une solution d'équilibre limite correspondant à $(\underline{Q}(\underline{\underline{\sigma}}),\ \underline{\dot{q}}(\underline{v}))$.

5.2. Théorème "d'unicité du champ de contrainte"

Sous les mêmes hypothèses on démontre également le théorème suivant:

Il peut exister plusieurs solutions d'équilibre limite correspondant à un même chargement limite $\underline{Q}$ (ou à une même vitesses de déformation $\underline{\dot{q}}$), mais il y a unicité du champ de contrainte dans la réunion des zones déformées de ces diverses solutions.

Il est clair que ce théorème développe une conséquence du théorème d'association du § 5.1.

On trouvera les démonstrations de ces résultats dans l'ouvrage cité de l'auteur.

6. CALCUL A LA RUPTURE

Comme nous y avons déjà insisté à plusieurs reprises, les théorèmes statique et cinématique et les méthodes associées, correspondent à des modes de raisonnement naturels. Ce sont les raisonnements utilisés en "calcul à la rupture":

- s'il est possible d'équilibrer un chargement donné par des forces intérieures admissibles, on affirme que le système résistera à ce chargement.
- s'il est possible de trouver pour un chargement donné, un mécanisme de ruine admissible tel que la puissance des forces motrices soit supérieure à la puissance des forces résistantes, on affirme que le système ne résistera pas à ce chargement.

Mais ce n'est que dans le cas du système standard que les propositions précedentes sont démontrées. Des contre-exemples ont été fournis qui montrent que dans le cas de systèmes non standards, ces propositions peuvent être contradictoires entre elles.

Ainsi l'utilisation courante des raisonnements du calcul à la rupture suppose que le système étudié est standard. Il ne faut pas chercher ailleurs les raisons des difficultés d'interprétation de certains calculs de stabilité en Mécanique des Sols.

BIBLIOGRAPHIE

Voir chap. V de l'ouvrage de l'auteur:
"Théorie de la Plasticité. . ."

J. MANDEL (1966), Mécanique des Milieux Continus, t. 2, Gauthier-Villars, Paris, 1966.

P.G. HODGE (1970), Limit analysis with multiple load parameters. Int.J. Solids & Structures, vol. 6, p. 661-675.

J. SALENÇON (1972), Charge limite d'un système non standard. in Plasticité et Viscoplasticité 1972, ed . D. Radenkovic et J. Salençon, Ediscience, Paris, 1974.

THEORIE DES CHARGES LIMITES POUR LES SYSTEMES NON-STANDARDS

1. SYSTEME NON-STANDARD

Il résulte de ce que nous avons dit au chapitre précédent qu'un système est non-standard, si le comportement plastique d'un de ses éléments ne satisfait pas les règles de convexité et normalité. En particulier, un système composé d'un matériau de Coulomb est en général non-standard, conformément à ce que nous avons dit à propos de la règle d'écoulement d'un tel matériau au chapitre 1. De même un système composé de plusieurs solides avec frottement de Coulomb aux interfaces est non-standard, que le matériau constitutif soit lui-même standard ou non-standard.

Pour un système non-standard les théorèmes statique et cinématique de la théorie classique des charges limites ne peuvent plus être établis. De plus des contre exemples ont été donnés montrant que les raisonnements intuitifs du calcul à la rupture peuvent dans ce cas conduire à des paradoxes.

L'étude de ce type de systèmes dans le but de mettre en évidence des théorèmes limites, évidemment moins puissants que les théorèmes classiques, qui leur soient applicables a été initiée par Drucker (1954) à propos des conditions aux interfaces; Radenkovic (1961, 1962) a engagé l'étude du point de vue plus général, suivi par de Josselin de Jong (1964, 1973), Palmer (1966), Sacchi et Save (1968), Collins (1968, 1969), Salençon (1969, 1972).

2. MATERIAU F

Soit M le matériau constitutif du système étudié: nous incluons dans M les interfaces éventuels entre solides constitutifs. On désigne par f le critère de plasticité de M.

Ainsi que nous l'avons dit à propos du comportement plastique des matériaux, et cela est également vrai pour les conditions de frottement aux interfaces, f est convexe en général. Nous poserons donc a priori que f est *convexe.*

Dire que le système est non standard c'est a dire qu'au moins en certains points la règle d'écoulement n'est *pas normale.*

Définissant comme au chapitre précédent dans le cas standard les champs de contraintes licites et les chargements licites ou S.P.A. pour le matériau M, il est clair par les mêmes arguments, que l'ensemble des chargements licites est un convexe K_M.

Les chargements limites pour le système (M), appartiennent nécessairement à K_M, mais faute de disposer de la propriété de normalité, on ne peut démontrer qu'ils sont sur la frontière de K_M.

$$Q_{lim} \in K_M \tag{2.1}$$

Désignons par F le matériau standard de critère f. Considérant alors le même système constitué de ce matériau, soit (F), il est évident que le convexe des chargements licites pour (F) est:

$$K_F \equiv K_M \tag{2.2}$$

et pour ce système (F) les chargements limites sont sur la frontière de K_F. (Ainsi on voit que le système en matériau standard va jusqu'au bout de ses capacités de résistance: le principe du travail maximal est bien un principe de bonne volonté du matériau).

La détermination de la frontière de $K_M \equiv K_F$, est malgré tout intéressante pour le système (M). En effet, même si l'on n'est pas assuré que le système (M) ira jusqu'au bout de ses possibilités, il est important de connaître quelles sont ces possibilités.

La détermination de la frontière de K_M se fera évidemment soit par la méthode statique sur le matériau F donc M, soit par la méthode *cinématique* pour le *matériau F* qui apparaît ainsi comme un *auxiliaire de calcul.* Cette dernière méthode est d'ailleurs la plus logique: la frontière de K_M constituant une limite extérieure pour l'ensemble des chargements limites, il convient de l'approcher elle-même par l'extérieur si l'on désire être certain de conserver le caractère extérieur de l'approximation proposée pour la borne des chargements limites de (M).

On a donc ainsi un théorème cinématique et une méthode cinématique pour le système (M):

tout chargement dont la puissance dans un champ de vitesses licite pour le système (F) est supérieure à la puissance dissipée dans ce champ de vitesse pour (F) est instable pour le système (M).

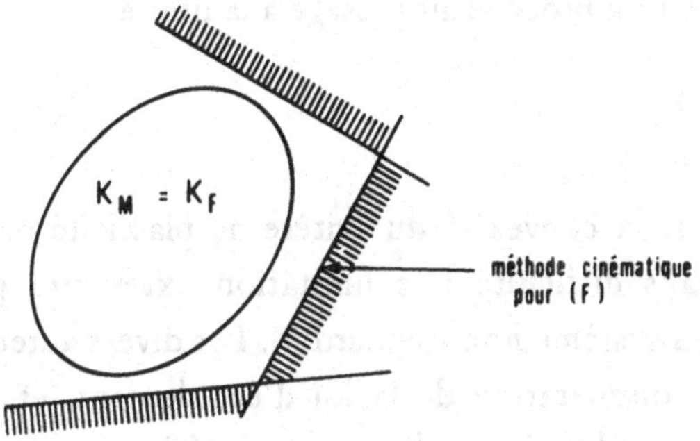

Fig. 1 – Méthode cinématique

3. REMARQUES

Il résulte de ce qui précède que dans le cas d'un système non-standard, l'appellation de "frontière d'écoulement" pour l'ensemble des chargements limites n'est pas justifiée et ne doit pas être retenue. On a affaire à un ensemble de chargements limites dont on sait seulement qu'il est borné extérieurement par la frontière d'écoulement du système standard de même critère.

Les implications possibles de cela sont:

• un même chargement pourrait être ou ne pas être limite suivant le trajet de charge suivi pour l'atteindre. Ceci signifierait que pour ce chargement il y aurait plusieurs solutions, les unes d'équilibre limite, les autres d'équilibre surabondant (c'est à dire ne permettant pas la déformation du système rigide plastique). La distinction chargement limite et chargement strictement stable ne serait plus possible à faire au niveau du système rigide plastique. Ce n'est qu'en revenant au matériau élasto-plastique et en suivant le trajet de charge que l'ambiguité serait évidemment levée.

A notre connaissance l'étude de ce problème n'a jusqu'à ce jour pas été poussée, et en particulier aucun exemple de tel chargement n'est connu.

• la signification même du matériau rigide plastique serait remise en cause:

le schéma rigide plastique n'ayant de signification que du point de vue des chargements limites, la non unicité des solutions rencontrée dans le cas du matériau standard n'avait aucune importance car l'unicité des chargements limites était, elle, assurée;

dans le cas présent, la non unicité des solutions rendant ambiguë la réponse sur les chargements limites, le système rigide plastique ne pourrait être défini intrinsèquement: il dépendrait en effet en particulier des propriétés élastiques du matériau à partir duquel on procède au passage à la limite.

4. *MATERIAU G*

En s'appuyant sur la convexité du critère de plasticité on a pu, comme nous l'avons vu, obtenir sans difficulté une limitation extérieure pour l'ensemble des chargements limites du système non standard M. Les divers auteurs cités plus haut se sont attachés, par la considération de la loi d'écoulement, et en faisant certaines hypothèses sur celle-ci, à obtenir une limitation intérieure pour ce même ensemble. Le but étant ainsi de définir une sorte de couronne, lieu possible des chargements limites du système M.

A notre connaissance tous les résultats obtenus dans ce domaine se ramènent au théorème énoncé par Radenkovic (1962) qui doit être exprimé sous forme suffisamment générale, en précisant au mieux les hypothèses nécessaires à son établissement.

4.1. Classe de matériaux M envisagée

L'idée de base est la notion de potentiel plastique.

On s'intérèsse aux matériaux non-standards dont le comportement peut être défini à partir de deux fonctions scalaires:

f critère,

g potentiel plastique,

fonctions convexes et telles que:

$$f(\underline{\underline{\sigma}}) \leqslant 0 \tag{4.1}$$

$$g(\underline{\underline{\sigma}}) \leqslant 0 \tag{4.2}$$

de la façon suivante:

$$\left\{\begin{array}{l} \cdot\ f(\underline{\underline{\sigma}}) < 0 \ \Rightarrow\ \underline{\underline{d}}^{p}(\underline{\underline{\sigma}}) = 0 \\[2ex] \cdot\ \forall \underline{\underline{\sigma}} \quad \text{vérifiant} \quad f(\underline{\underline{\sigma}}) = 0 \ \Rightarrow\ \exists\, \underline{\underline{\sigma}}'(\underline{\underline{\sigma}}) \\ \qquad\qquad\qquad\qquad\qquad \text{vérifiant :} \quad g(\underline{\underline{\sigma}}') = 0 \\ \text{et tel que :}\ \ \underline{\underline{d}}^{p}(\underline{\underline{\sigma}}) \in \lambda\, \partial g(\underline{\underline{\sigma}}')\ , \quad \lambda \geqslant 0 \\ \qquad\quad \text{et} \quad \mathrm{tr}\,[\,(\underline{\underline{\sigma}} - \underline{\underline{\sigma}}') \cdot \underline{\underline{d}}^{p}\,(\underline{\underline{\sigma}})] \geqslant 0\ . \end{array}\right. \tag{4.3}$$

On remarque que:

• les matériaux standards sont un cas particulier de ce type de matériaux non-standards. Pour eux il suffit de prendre $g = f$.

• pour un tel matériau non-standard, la fonction g n'est pas définie de manière unique de même que la correspondance $\underline{\underline{\sigma}} \to \underline{\underline{\sigma}}'(\underline{\underline{\sigma}})$.

Ainsi, par exemple, si g vérifie (4.1, 4.2, 4.3), avec la correspondance $\underline{\underline{\sigma}} \to \underline{\underline{\sigma}}'$, alors la fonction g_m définie par:

$$g_m(\underline{\underline{\alpha}}) = g(m\underline{\underline{\alpha}}), \quad \forall \underline{\underline{\alpha}}, \quad m > 1 \tag{4.4}$$

et la correspondance $\underline{\underline{\sigma}} \to \underline{\underline{\sigma}}'' = \underline{\underline{\sigma}}'/m$

vérifient aussi (4.1, 4.2, 4.3) et permettent également de définir la loi de comportement du même matériau.

La transformation de la fonction g exprimée par (4.4), (homothétie) n'est évidemment pas la seule à posséder cette propriété; on peut également effectuer des translations (dans certaines limites). etc. . . .

•la dernière inégalité intervenant dans (4.3) ainsi que la convexité de f et g impliquent que le domaine $g(\underline{\underline{\sigma}}) < 0$ est intérieur au domaine $f(\underline{\underline{\sigma}}) < 0$, soit:

$$g(\underline{\underline{\sigma}}) \leqslant 0 \Rightarrow f(\underline{\underline{\sigma}}) \leqslant 0 \tag{4.5}$$

4.2. Matériau G

Pour un matériau non-standard M du type ci-dessus, et pour chaque fonction g, on définit le matériau G associé: matériau *standard* dont le critère de plasticité est g.

4.3. Théorème de Radenkovic

On considère le système (G) homologue du système (M) et constitué du matériau rigide-plastique G.

Il résulte de (4.5) que l'ensemble K_G des chargements licites pour (G) a la propriété:

$$K_G \subset K_M \equiv K_F . \tag{4.6}$$

On démontre le théorème suivant:

tout chargement limite pour le système (M) n'est pas intérieur à K_G.

Ainsi le chargements limites du système (M) se trouvent nécessairement dans la "couronne" située entre K_G et K_F , frontières comprises.

Le système (G) étant un système standard, la détermination de sa frontière d'écoulement, frontière de K_G, peut se faire par les méthodes statique et cinématique de la théorie classique.

Puisque la frontière de K_G constitue une limite intérieure pour l'ensemble des chargements limites de (M), il convient de l'approcher elle-même par l'intérieur si l'on désire être certain de conserver le caractère intérieur de l'approximation proposée pour la borne des chargements limites de (M).

On a donc ainsi un théorème statique et une méthode statique pour le système (M):

tout chargement stable pour le système (G) est stable pour le système (M).

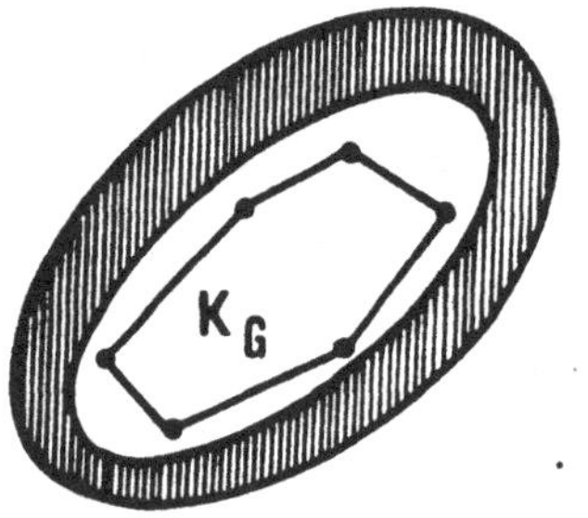

Fig. 2 – Méthode statique

4.4. Remarques sur l'application de ce théorème

• Ainsi qu'on l'a dit au § 4.1, il n'y a pas unicité de la fonction g intervenant dans la définition de la loi de comportement de (M). Le théorème énoncé ci-dessus est valable quelle que soit g. Il est évident que le résultat obtenu sera d'autant plus puissant que le domaine K_G sera le plus étendu.

• Les domaines K_G correspondant aux diverses fonctions g possibles ne sont pas nécessairement emboités: (4.6) étant valable quelle que soit g, il convient d'écrire le théorème sous la forme:

$$\bigcup_{\forall g} K_g \subset K_M \equiv K_F \qquad (4.7)$$

et les chargements limites pour (M) sont extérieurs à $\bigcup_{\forall g} K_G$.

4.5. Une variante possible

Le théorème énoncé, et les conditions imposées à g pour sa démonstration, sont indépendants de la nature du système étudié et du processus de chargement qu'il subit; partant d'une propriété de la loi de comportement de M, on a énoncé un théorème général pour tout système (M) constitué de ce matériau. soumis à n'importe quel processus de chargement.

De Josselin de Jong (1964, 1973), reprenant la démonstration du théorème, remarque que dans chaque cas, pour chaque problème étudié, les conditions sur g sont moins restrictives:

en effet, on peut démontrer qu'un chargement limite Q_1 pour (M) n'est pas à l'intérieur de K_G ,dès que la fonction g vérifie (4.3), et en particulier la dernière inégalité, pour un champ $\underline{\underline{\sigma}}$ d'une solution d'équilibre limite correspondant à Q_1 . On obtient donc ainsi un théorème plus fin que le précédent puisque la fonction g à faire intervenir, qui peut varier à l'intérieur du système, est soumise à une condition plus large que (4.3).

L'application de ce théorème sous cette forme est évidemment impossible puisqu'elle suppose connus les chargements limites et les solutions correspondantes. Par contre, une forme intermédiaire entre le théorème du § 4.3 et celui énoncé ci-dessus, consisterait à essayer d'estimer à l'avance, en chaque point du système, le domaine dans lequel risque de se trouver la contrainte $\underline{\underline{\sigma}}$ pour l'équilibre limite dans une sollicitation d'une type donné, et à imposer à g de vérifier (4.3) pour les $\underline{\underline{\sigma}}$ dans ce domaine.

4.6. Exemples de tels matériaux non-standards

En Mécanique des Sols, les matériaux non-standards rencontrés sont essentiellement des matériaux de Coulomb, $\varphi \neq 0$. Avec les règles d'écoulement proposées au chapitre 1, basées sur la notion d'angle de dilatance, ces matériaux ressortissent au type du § 4.1. On a alors pour f un critère de Coulomb d'angle φ et pour g un potentiel plastique de Coulomb d'angle $0 \leqslant \nu \leqslant \varphi$.

Ainsi le matériau F et le matériau G sont dans ce cas deux matériaux de Coulomb standards, d'angles de frottement interne φ et ν .

Si on applique le théorème sous sa forme générale avec les hypothèses du § 5.1, on prendra pour G la cohésion:

(4.8) $$C_g = C \operatorname{tg} \nu / \operatorname{tg} \varphi$$

en désignant par C la cohésion du matériau M; en effet la fonction g qui correspond au plus grand K_G, sans tenir compte des particularités du problème, est telle que les deux matériaux aient même pression de cohésion $H = C \cot g\, \varphi = C_g \cot g\, \nu$.

On remarque que si la déformation plastique s'effectue sans variation de volume, $\nu = 0$, ceci conduit pour le matériau G au comportement d'un liquide:

$$\nu = 0, \qquad C_g = 0\,!$$

La limite inférieure ainsi obtenue pour les chargements limites de (M) n'offrira dans ce cas pas grand intérêt. On comprend alors pourquoi la seconde variante proposée au § 4.5 est, lorsqu'il est possible de l'appliquer, de la plus grande importance.

En ce qui concerne les interfaces, à propos desquels on trouvera une discussion détaillée dans l'ouvrage cité de l'auteur, l'interface lisse est standard, l'interface à frottement sec quelconque ne l'est pas (ceci revient à définir la "meilleure" fonction g par une propriété d'enveloppe, comme indiqué par Palmer).

5. *SOLUTIONS POSSIBLES*

On parvient pour certains problèmes, pour un système non-standard, à construire des "solutions d'équilibre limite possibles", c'est à dire un champ de contrainte $\underline{\underline{\sigma}}$ licite

un champ de vitesse $\underline{v}$ licite

et associés pour la loi de comportement non-standard.

Dans ce cas, on peut dire que le chargement $\underline{Q}$ correspondant à cette solution, est un chargement limite possible. En particulier, si l'on a affaire à un matériau du type étudié au § 5., cela signifie que $\underline{Q}$ est dans la couronne comprise entre K_G et K_F frontières comprises.

Mais, conformément à ce que l'on a dit au § 3, on n'est pas assuré qu'il n'existe pas pour ce même chargement des solutions d'équilibre surabondant (c'est à dire où le champ $\underline{v}$ associé ne puisse être que nul).

BIBLIOGRAPHIE

Voir chapitre 5 de l'ouvrage de l'auteur *Théorie de la Plasticité* . . .

I.F. COLLINS (1969), The upper bound theorem for rigid plastic solids generalized to include Coulomb friction. J. Mech. Phys. Solids, vol. 17, p. 323-338.

D.C. DRUCKER (1954), Coulomb friction, Plasticity and limit loads. J. Appl. Mech. Trans. ASME, vol. 21, p. 71-74.

G. de JOSSELIN de JONG (1964), Lower bound collapse theorem and lack of normality of strain rate to yield surface for soils. Proc. IUTAM Symp. on Rheology and Soil Mech., Grenoble, p. 69-75.

G. de JOSSELIN de JONG (1973), A limit theorem for material with internal friction. Proc. Symp. on the Role of Plasticity in Soil Mechanics, Cambridge (G.B.), p. 12-21.

A.C. PALMER (1966), A limit theorem for materials with non associated flow laws. J. Mécanique, vol. 5, n. 2, p. 217-222.

D. RADENKOVIC (1961), Théorèmes limites pour un matériau de Coulomb à dilatation non-standardisée. C.R. Ac. Sc. Paris, 252, p. 4103-4104.

D. RAKENKOVIC (1962), Théorie des charges limites. Séminaire de Plasticité, ed. J. Mandel, Paris, p. 129-142.

G. SACCHI et M. SAVE (1968), A note on the limit loads of non-standard materials. Meccanica, vol. 3, n. 1, p. 43-45.

J. SALENÇON (1972), Charge limite d'un système non-standard – Un contre-exemple pour la théorie classique. Séminaire Plasticité et Viscoplasticité 1972, ed. D. Radenkovic et J. Salençon, Ediscience, Paris, 1974.

J. SALENÇON (1969), La théorie des charges limites dans la résolution des problèmes de plasticité en déformation plane. Thèse, Doct. es Sc. Paris.

EQUILIBRES LIMITES PLANS ET DE REVOLUTION

1. INTRODUCTION

Les problèmes d'équilibre limite en déformation plane possèdent certaines propriétés mathématiques qui simplifient énormément leur résolution en particulier du point de vue numérique (emploi de la méthode des caractéristiques).

C'est d'ailleurs pour cette raison que nombre de problèmes fondamentaux de la Mécanique des Sols ont été ramenés à ce type de problèmes qui en permettait l'étude du point de vue théorique: par exemple, capacité portante des fondations assimilées à des semelles filantes,. . .

Les problèmes de symétrie axiale sous certaines hypothèses (critère de type courbe intrinsèque et hypothèse de Haar-Karman) se résolvent de manière semblable.

Nous ne donnerons pas ici une étude détaillée de ces problèmes. On pourra pour cela se reporter aux ouvrages classiques spécialisés et à l'ouvrage déjà cité de l'auteur. Nous nous bornerons à insister sur les idées directrices et à rappeler les résultats fondamentaux, dans le cas du matériau isotrope.

2. PROBLEMES DE DEFORMATION PLANE

2.1. Le critère de plasticité dans le plan

• On démontre que si l'on a affaire à un matériau *standard* problèmes de déformation plane, à un critère courbe intrinsèque dans le plan de la déformation.

• Dans le cas d'un matériau non-standard:

si le critère du matériau est du type courbe intrinsèque (par ex.: critère de Coulomb), le critère de plasticité se ramène dans le plan à un critère de type courbe intrinsèque, dès que la loi de comportement implique que le plan de la déformation contient les contraintes principales extrêmes. Il en est ainsi par exemple si la déformation plastique dérive d'un potentiel de Mises, ou de Tresca ou de Coulomb, comme indiqué au chapitre 1.

si le critère de plasticité du matériau est quelconque et que la déformation plastique dérive d'un potentiel de Mises, il en va encore de même: le critère de plasticité s'exprime encore dans le plan sous la forme d'un critère courbe intrinsèque.

2.2. Découplage des problèmes en contrainte et en vitesse

Ainsi, dans les cas indiqués ci-dessus, et qui recouvrent presque tous les problèmes pratiques, on a dans les zones plastiques des solutions d'équilibre limite une relation entre les contraintes dans le plan (x,y) de la déformation, de type courbe intrinsèque, qui exprime la condition de plasticité.

Compte-tenu d'autre part de l'isotrope, les composantes σ_{xz} et σ_{yz} du tenseur des contraintes sont nulles.

On voit alors que l'on dispose de 3 équations qui ne font intervenir que les trois composantes de $\underline{\underline{\sigma}}$ dans le plan:

le problème pour les contraintes dans les zones plastiques est donc séparé de celui des vitesses. Celui-ci se résoudra ensuite, une fois déterminé le champ de contraintes dans les zones plastiques.

Les paragraphes suivants rappellent les résultats obtenus pour les problèmes en contraintes et en vitesses.

2.3. Déformation plane – Matériau non-homogène – Problème pour les contraintes

• Notations:

contraintes principales $\sigma_1 \geqslant \sigma_2$ dans le plan

$p = -(\sigma_1 + \sigma_2)/2, \quad \theta = (Ox, \sigma_1)$

$R = (\sigma_1 - \sigma_2)/2$

$\sigma_x = -p + R\cos 2\theta$

$\sigma_y = -p - R\cos 2\theta$

$\sigma_{xv} = R \sin 2\theta$

• Courbe intrinsèque, matériau non-homogène:

$$R = R(p, x, y)$$

$$\partial R/\partial p = \sin\varphi \;, \qquad \varphi(p, x, y) \,.$$

• Lignes caractéristiques:

$$(2.1) \qquad \frac{dy}{dx} = \begin{cases} \operatorname{tg}[\theta + (\frac{\pi}{4} + \frac{\varphi}{2})] \,, & \text{ligne } \beta \\ \operatorname{tg}[\theta - (\frac{\pi}{4} + \frac{\varphi}{2})] \,, & \text{ligne } \alpha \,. \end{cases}$$

• Relations le long des caractéristiques:

$$\begin{cases} dp + \dfrac{2R}{\cos\varphi}\, d\theta - (\rho F^{\alpha} + \dfrac{\partial R}{\partial x_{\beta}})\, dx^{\alpha} = 0\,, & \text{ligne } \alpha \\ dp - \dfrac{2R}{\cos\varphi}\, d\theta - (\rho F^{\beta} + \dfrac{\partial R}{\partial x_{\alpha}})\, dx^{\beta} = 0\,, & \text{ligne } \beta\,. \end{cases} \tag{2.2}$$

R et φ fonctions de (p, x, y),

F : force de masse

composantes covariantes et contravariantes dans la base normée tangente aux lignes caractéristiques.

2.4. Déformation plane – Matériau non-homogène, standard. Problème pour les vitesses

• Notations:

v_{α}, v_{β} composantes covariantes de la vitesse dans la base normée tangente aux lignes caractéristiques.

• Lignes caractéristiques pour les vitesses ≡ lignes caractéristiques pour les contraintes.

• Relations le long des caractéristiques:

$$\begin{cases} D_{\alpha} v_{\alpha} = \partial_{\alpha} v_{\alpha} - v_{\alpha}\, \mathrm{tg}\,\varphi\, \partial_{\alpha}(\theta - \frac{\varphi}{2}) - v_{\beta} \dfrac{1}{\cos\varphi}\, \partial_{\alpha}(\theta - \frac{\varphi}{2}) = 0\,, & \text{ligne } \alpha\,. \\ D_{\beta} v_{\beta} = \partial_{\beta} v_{\beta} + v_{\alpha} \dfrac{1}{\cos\varphi}\, \partial_{\beta}(\theta + \frac{\varphi}{2}) + v_{\beta}\, \mathrm{tg}\,\varphi\, \partial_{\beta}(\theta + \frac{\varphi}{2}) = 0, & \text{ligne } \beta\,. \end{cases} \tag{2.3}$$

Les lignes α et β sont lignes d'extension nulle.

• Condition de positivité de λ ($\underline{\underline{d}}^{p} \in \lambda\,\partial f$, $\lambda \geqslant 0$) :

$$\begin{aligned} D_{\alpha} v_{\beta} + D_{\beta} v_{\alpha} = {} & \partial_{\alpha} v_{\beta} + \partial_{\beta} v_{\alpha} - (v_{\alpha}\, \mathrm{tg}\,\varphi + v_{\beta} \frac{1}{\cos\varphi})\, \partial_{\beta}(\theta - \frac{\varphi}{2}) + \\ & + (v_{\alpha} \frac{1}{\cos\varphi} + v_{\beta}\, \mathrm{tg}\,\varphi)\, \partial_{\alpha}(\theta + \frac{\varphi}{2}) \geqslant 0\,. \end{aligned} \tag{2.4}$$

Cette condition peut être vérifiée par construction de l'hodographe et étude de son orientation.

• Solution faibles : discontinuité de la vitesse

Lignes de discontinuité = lignes caractéristiques.

Franchissement d'une ligne α :

$$(2.5)\qquad \begin{cases} [\![v_\beta]\!] = 0 \\ [\![v_\alpha]\!] \geqslant 0 \\ D_\alpha \, [\![v_\alpha]\!] = 0 \,. \end{cases}$$

Franchissement d'une ligne β :

$$(2.6)\qquad \begin{cases} [\![v_\alpha]\!] = 0 \\ [\![v_\beta]\!] \geqslant 0 \\ D_\beta \, [\![v_\beta]\!] = 0 \,. \end{cases}$$

3. *PROBLEMES EN SYMETRIE AXIALE*

3.1. *Hypothèse de Haar-Karman*

Soient r, ω, z les coordonnées cylindriques.

On considère les problèmes d'écoulement plastique libre pour lesquels la distribution des contraintes a les propriétés suivantes:

• symétrie axiale autour de Oz : $\underline{\underline{\sigma}}$ indépendent de ω

• $\sigma_{\omega\omega}$ = contrainte principale.

Le matériau constitutif du système est isotrope, obéit à un critère de plasticité du type courbe intrinsèque, éventuellement non-homogène en r et z mais indépendent de ω.

L'hypothèse de Haar-Karman consiste à poser que le régime d'écoulement en zone plastique est un régime d'arête, σ_ω étant égale à l'une des contraintes principales dans le plan éridien. Les motivations de cette hypothèse ont été exposées par Shield (1955), Cox, Eason et Hopkins (1961).

L'intérêt pratique en est que le problème pour les contraintes dans les zones plastiques peut alors, comme dans le cas plan, être résolu indépendamment de celui des vitesses; le champ des vitesses étant ensuite déterminé une fois les contraintes connues.

Les paragraphes suivants rappellent les principaux résultats relatifs à ces deux problèmes.

3.2. Symétrie axiale – Matériau non-homogène – Problème pour les contraintes

· Notations

contraintes principales dans le plan méridien:

$$p = -(\sigma_1 + \sigma_3)/2$$

$$R = (\sigma_1 - \sigma_3)/2$$

$$\theta = (\mathrm{Or}, \sigma_1)$$

· Hypthèse de Haar-Karman:

$$\sigma_\omega = \sigma_2 \quad \text{et :} \quad \sigma_2 = \sigma_1 \quad \text{ou} \quad \sigma_2 = \sigma_3$$

$$\sigma_r = -p + R\cos 2\theta$$

$$\sigma_z = -p - R\cos 2\theta$$

$$\sigma_{rz} = R\sin 2\theta$$

$$\sigma_\omega = -p - \epsilon R \qquad \begin{array}{l} \epsilon = -1 \text{ correspond à } \sigma_\omega = \sigma_1 \\ \epsilon = +1 \text{ correspond à } \sigma_\omega = \sigma_3 \end{array}$$

· Courbe intrinsèque, matériau non-homogène:

$$R = R(p, r, z)$$

$$\partial R/\partial p = \sin\varphi(p, r, z)$$

· Lignes caractéristiques:

$$\frac{dz}{dr} = \begin{cases} \mathrm{tg}[\theta + (\frac{\pi}{4} + \frac{\varphi}{2})] & \text{, ligne } \beta \\ \mathrm{tg}[\theta - (\frac{\pi}{4} + \frac{\varphi}{2})] & \text{, ligne } \alpha\,. \end{cases} \tag{3.1}$$

• Relations le long des caractéristiques:

$$
(3.2)\quad \begin{cases}
dp + 2\dfrac{R}{\cos\varphi}\,d\theta - \{\rho F^{\alpha} + \dfrac{\partial R}{\partial x_{\beta}} - \dfrac{R}{r\cos\varphi}\,[\sin(\theta - (\dfrac{\pi}{4} + \dfrac{\varphi}{2}))\\
\qquad - \epsilon \sin(\theta + (\dfrac{\pi}{4} + \dfrac{\varphi}{2}))]\}\,dx^{\alpha} = 0, \quad \text{ligne } \alpha\\
dp - 2\dfrac{R}{\cos\varphi}\,d\theta - \{\rho F^{\beta} + \dfrac{\partial R}{\partial x_{\alpha}} + \dfrac{R}{r\cos\varphi}\,[\sin(\theta + (\dfrac{\pi}{4} + \dfrac{\varphi}{2}))\\
\qquad - \epsilon \sin(\theta - (\dfrac{\pi}{4} + \dfrac{\varphi}{2}))]\}\,dx^{\beta} = 0, \quad \text{ligne } \beta .
\end{cases}
$$

3.3. Symétrie axiale – Matériau non-homogène, standard. Problème pour les vitesses

• Notations:

v_r, v_ω, v_z composantes covariantes de la vitesse dans le plan méridien dans la base tangente aux lignes caractéristiques.

• Lignes caractéristiques pour les vitesses = lignes caractéristiques pour les contraintes.

• Relations le long des caractéristiques:

$$
(3.3)\quad \begin{cases}
D_{\alpha} v_{\alpha} + \dfrac{v_r}{2r}(1 + \epsilon \sin\varphi) = 0, & \text{ligne } \alpha .\\
D_{\beta} v_{\beta} + \dfrac{v_r}{2r}(1 + \epsilon \sin\varphi) = 0, & \text{ligne } \beta .
\end{cases}
$$

• Condition de positivité de λ et μ :

$$
(3.4)\quad \begin{cases}
\epsilon \dfrac{v_r}{r} \leqslant 0\\
D_{\alpha} v_{\beta} + D_{\beta} v_{\alpha} \geqslant - \epsilon \dfrac{v_r}{r}(1 + \epsilon \sin\varphi) .
\end{cases}
$$

• Solutions faibles: discontinuité de la vitesse.

Lignes de discontinuité = lignes caractéristiques

Franchissement d'une ligne α :

$$[\![v_\beta]\!] = 0$$

$$[\![v_\alpha]\!] \geqslant 0$$

$$D_\alpha [\![v_\alpha]\!] = -\frac{1}{2r} [\![v_\alpha]\!] \cos(\theta - (\frac{\pi}{4} + \frac{\varphi}{2})) . (1 + \epsilon \sin\varphi)$$

de même au franchissement d'une ligne β.

4. PROBLEME POUR LES VITESSES POUR LE MATERIAU NON-STANDARD

Les résultats donnés aux § 2.4 et 3.3 dans le cas du matériau standard permettent aussi de résoudre le problème de la détermination des vitesses pour les systèmes constitués de certains matériaux non-standards.

C'est en particulier le cas pour les matériaux de Coulomb (φ) dont la règle d'écoulement dérive d'un potentiel plastique, de Coulomb également (ν).

On procédera de la façon suivante:

le champ de contraintes dans les zones plastiques étant déterminé en utilisant les résultats des § 2.3 ou 3.2, on passera au problème pour les vitesses.

Dans le cas de la déformation plane par exemple, on remarque que

d'après la loi de comportement, les directions principales de $\underline{\underline{d}}$ sont connues (≡ celles de $\underline{\underline{\sigma}}$)

et l'on doit avoir:

$$d_1^p = \lambda(1 + \sin\nu) , \qquad d_2^p = -\lambda(1 - \sin\nu) , \qquad \lambda \geqslant 0 ,$$

ce qui signifie que $\underline{\underline{d}}$ doit admettre pour directions d'extension nulle les lignes γ et δ inclinées à $(\pi/4 + \nu/2)$ sur σ_1.

On a donc ainsi les deux familles de lignes γ et δ connues dans les zones plastiques, le long desquelles ont lieu les relations homologues de (2.3) où l'on doit remplacer α , β par γ , δ et φ par ν . Ceci permet donc de déterminer le champ des vitesses. La condition $\lambda \geqslant 0$, de la relation de comportement (4.1), conduit de même à l'inégalité homologue de (2.4) par la même substitution $\alpha, \beta, \varphi \Rightarrow$

$\Rightarrow \gamma, \delta, \nu$.

5. *SIGNIFICATION DES SOLUTIONS AINSI CONSTRUITES*

S'agissant du matériau rigide plastique, les solutions construites en déformation plane ou en symétrie axiale en s'appuyant sur les propriétés indiquées ci-dessus, doivent être interprétées dans le cadre de la théorie des charges limites, qui seule permet d'en dégager la signification. Le travail de Bishop (1953) est à notre connaissance un des premiers où ce probléme ait été clairement posé.

5.1. *Cas du matériau standard*

a) La majorité des solutions proposées pour les problèmes plans consiste en:

un champ de contrainte $\underline{\underline{\sigma}}$ en équilibre et à la limite d'écoulement déterminé dans un partie *P* du système étudié

un champ de vitesse cinématiquement et plastiquement admissible, déterminé dans tout le système, associé à $\underline{\underline{\sigma}}$ par la loi de comportement dans *P*, et pour lequel l'extérieur de *P* ne subit que des mouvements rigidifiants.

Une telle solution est dite incomplète au sens de Bishop, ressortit à l'approche cinématique, et on démontre que les équations de Kötter, (2.2) permettent une application commode de la méthode cinématique.

b) Certaines solutions en déformation plane, et bon nombre en symétrie axiale, ne fournissent que le champ de contrainte $\underline{\underline{\sigma}}$ ci dessus dans une zone *P*, sans champ de vitesse associé.

Ces solutions ne sont en toute rigueur pas interprétables, et ne peuvent fournir que des indications "sentimentales".

c) Les résultats des § 2.3 ou 3.2 peuvent être utilisés pour la construction de solutions purement statiques, fournissant un champ de contraintes licite à la limite d'écoulement dans un partie du système et un prolongement licite de ce champ dans le reste du système. Le chargement correspondant est un chargement licite.

d) Les résultats du § 2.4 peuvent être utilisés pour la construction des solutions purement cinématiques, indépendamment de la construction de champs de containtes associés aux champs de vitesses.

Ces solutions seront interprétées au moyen du théorème cinématique sous forme classique.

e) Enfin, certaines solutions proposées fournissent à la fois un champ de contraintes licite dans tout le système et un champ de vitesses licite associé. Ce sont des solutions complètes.

Une telle solution correspond donc à un chargement limite.

5.2. Cas du matériau non-standard

a) La majorité des solutions construites dans le cas du matériau non-standard correspond au cas b) du § 5.1. C'est dire qu'elles sont, en toute rigueur, impossibles à interpréter.

C'est en particulier le cas pour les solutions construites en Mécanique des Sols pour le matériau de Coulomb. Les rares auteurs qui n'aient pas fait le silence sur le problème de la signification du résultat fourni par une telle solution, ont le plus souvent proposé de faire l'interprétation en se référant à la solution homologue pour le matériau de Tresca.

b) Beaucoup des solutions ci-dessus peuvent en fait être associées cinématiquement à un champ de vitesse licite pour le matériau Standard F correspondant. On se trouve donc alors dans le cas de l'application du théorème cinématique pour le matériau non standard.

c) On possède aussi pour certains problèmes des solutions fournissant un champ de contrainte licite pour le matériau non-standard M. Les théorèmes de l'analyse limite dans le cas des matériaux non-standards ne permettent pas d'interpréter ce type de solutions: on peut seulement affirmer que les chargements correspondants sont licites pour le matériau F.

d) Enfin quelques solutions ont été proposées qui associent un champ de contraintes licite dans une zone *P* à un champ de vitesses licite dans tout le système, pour le matériau non-standard. Ces solutions apparentées aux solutions incomplètes du § 5.1 (a), ne peuvent, pas plus que celles des § 5.1 (b) et 5.2 (a), recevoir une interprétation autre que "sentimentale".

e) Des solutions statiques pour le matériau G peuvent être construites et correspondent alors à l'approche statique pour le matériau non-standard.

Il résulte de ce qui précède qu'en fait les calculs faits dans le cas du matériau de Coulomb par le mécanicien des sols, même lorsque le problème des vitesses n'est pas évoqué, reposent sur l'hypothèse implicite que l'on travaille avec le matériau standard F: on se place dans le cas 5.2 (b), admettant, sans le construire, que le champ de vitesses associé existe comme dans le cas du matériau de Tresca standard.

Le matériau de Coulomb standard est l'outil de travail implicite employé pour nombre de calculs de capacités portantes par exemple. Il convient de signaler que

Davis et Booker (1971) ont procédé à la construction de solutions du type 5.2 (d), pour un problème de capacité portante –sous réserve d'une vérification éventuellement plus approfondie de la condition $\lambda \geqslant 0$ de (4.1)–: les capacités portantes obtenues ne subissaient que des variations très peu sensibles suivant les valeurs de ν, $(0 \leqslant \nu \leqslant \varphi)$.

BIBLIOGRAPHIE

Voir le chapitre IV de l'ouvrage de l'auteur *Théorie de la Plasticité...*

B.G. BEREZANCEW (1952), Problème de l'équilibre limite d'un milieu pulvérulent en symétrie axiale. Ed. Litt. Tech. Theor. Moscou.

J.F. BISHOP (1953), On the complete solution to problem of deformations of a plastic rigid material. J. Mech. Phys. Solids, vol. 2, p. 43-53.

A.D. COX, G. EASON & H.G. HOPKINS (1961), Axially symmetric plastic deformations in soils. Phil. Trans. Roy. Soc. London, A, 1036, 254, p. 1–45.

E.H. DAVIS & J.R. BOOKER (1971), The bearing capacity of strip footings from the standpoint of plasticity theory. Research report, Civil Eng. Lab., Univ. Sydney.

A. HAAR et Th. KARMAN (1909), Zur theorie der Spannungzustände in plastischen und sandartigen Medien. Nach. Ges. Wiss. Göttingen, Math. Phys. Kl., p. 204-218.

R. HILL (1950), The mathematical theory of Plasticity, Clarendon Press, Oxford, 1950.

J. MANDEL (1942), Equilibre par tranches planes des solides à la limite d'écoulement. Thèse, ed. Louis Jean, Gap et Travaux, juin, juillet, décembre 1943.

J. SALENÇON (1969), La théorie des charges limites dans la résolution des problèmes de plasticité en déformation plane. Thèse Doct. es Sc., Paris.

R.T. SHIELD (1955), On the plastic flow of metals under conditions of axial symmetry. Proc. Proc. Roy. Soc., 233, A, 1183, p. 267-287.

V.V. SOKOLOVSKI (1960), Statics of soil media. Butterworths, Sc. Publ. Londres.

V.V. SOKOLOVSKI (1965). Statics of granular media. Pergamon Press.

EXEMPLES D'APPLICATION

Ce chapitre présente quelques exemples d'application des méthodes de calcul pour le matériau rigide plastique, standard ou non-standard évoquées précédemment.

1. UN CALCUL DE CAPACITE PORTANTE

Les résultats présentés dans ce paragraphe sont extraits d'un travail réalisé par P. Florentin et Y. Gabriel sous la direction de l'auteur.

On désire calculer la capacité portante d'une fondation filante superficielle sur un sol de Coulomb non homogène: angle de frottement interne φ et cohésion variable linéairement avec la prondeur, C.

On utilise la méthode de calcul classique basée sur la théorie des équilibres limites plans, considérée uniquement du point de vue des contraintes. La solution en contraintes dans la zone plastique est construite par la méthode des caractéristiques.

(En fait, sans que cela ait été vérifié, il est vraisemblable qu'il est possible d'associer un champ de vitesses à ce champ de contraintes en faisant l'hypothèse du matériau standard ⇒ approche cinématique pour le matériau non-standard).

L'analyse préliminaire précise du problème montre qu'il est possible de grouper entre eux les divers paramètres intervenant dans ce problème (utilisation de l'analyse dimensionnelle, du théorème des états correspondants, et examen des équations); on réduit ainsi le nombre de paramètres et le problème se trouve ramené à celui de la capacité portante d'une fondation sur sol homogène, (fig. 2).

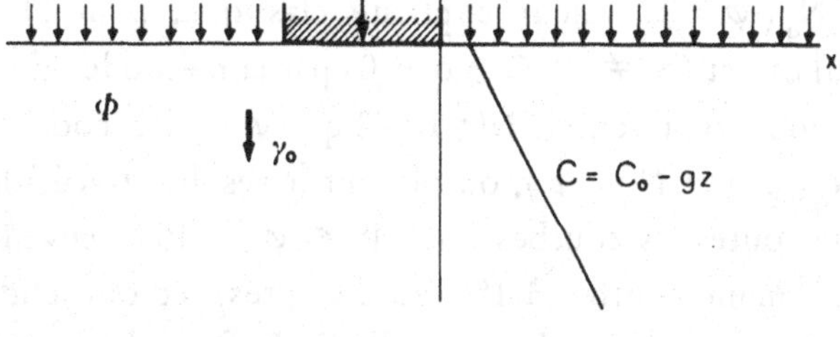

Fig. 1 – Capacité portante d'une fondation sur sol non-homogène (notations)

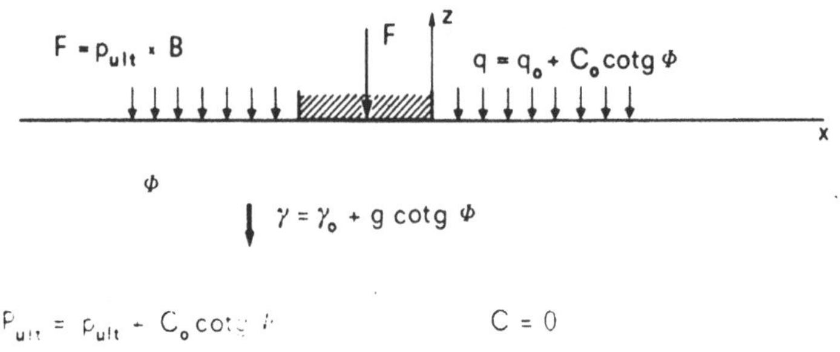

Fig. 2

Pour celui-ci, renonçant à la méthode dite "de superposition", et à la formule linéaire qui lui correspond on a procédé à la construction de la solution globale, au moyen des équations de l'équilibre limite plan. On obtient ainsi pour chaque valeur de φ une courbe:

$$p^{o}_{ult} + C_o \cot g \varphi = (\rho_o + C_o \cot g \varphi) \cdot N\left(\frac{\gamma_o + g \cot \varphi}{g_o + C_o \cot g \varphi} \cdot \frac{B}{2}, \varphi\right)$$

qui inclut l'effet de tous les paramètres. (Le calcul a été mené dans le cas de la fondation parfaitement rugueuse).

Les coefficients classiques N_q et N_γ, correspondent ici à:

$$N_q(\varphi) = N(0, \varphi)$$

$$N_\gamma(\varphi) = \lim_{\gamma B/2q \to \infty} \cdot \frac{2q}{\gamma B} N\left(\frac{\gamma B}{2q}, \varphi\right),$$

en retrouve pour $N_q(\varphi)$ la valeur explicite classique, pour $N_\gamma(\varphi)$ la valeur obtenue par le calcul direct ($\gamma \neq 0, C = q = 0$) par la méthode "du tir".

En utilisant, pour représenter $N(\gamma B/2q, \varphi)$, des coordonnées réduites faisant intervenir $N_q(\varphi)$ et $N_\gamma(\varphi)$, on obtient le résultat tracé à la figure 4, où il est remarquable que toutes les courbes pour $4^\circ \leqslant \varphi \leqslant 40^\circ$ peuvent pratiquement être assimilées à une même courbe (à 1% ou 2% près), ce qui rend l'utilisation de ces résultats pratiquement aussi aisée que celle de la formule de superposition. La figure 3 montre un exemple de réseau de caractéristiques correspondant à un tel calcul.

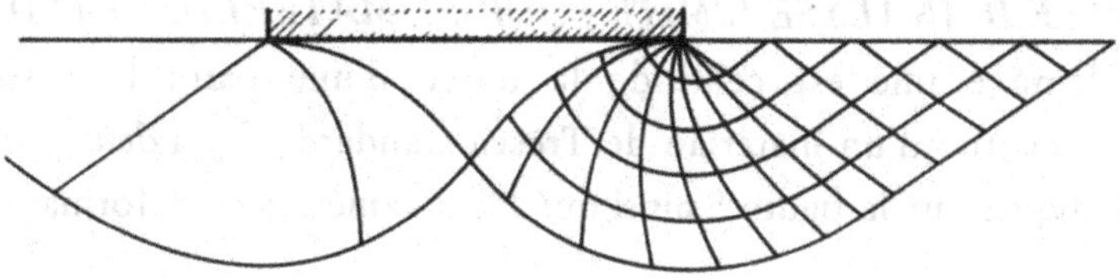

Fig. 3·

Fig. 3·

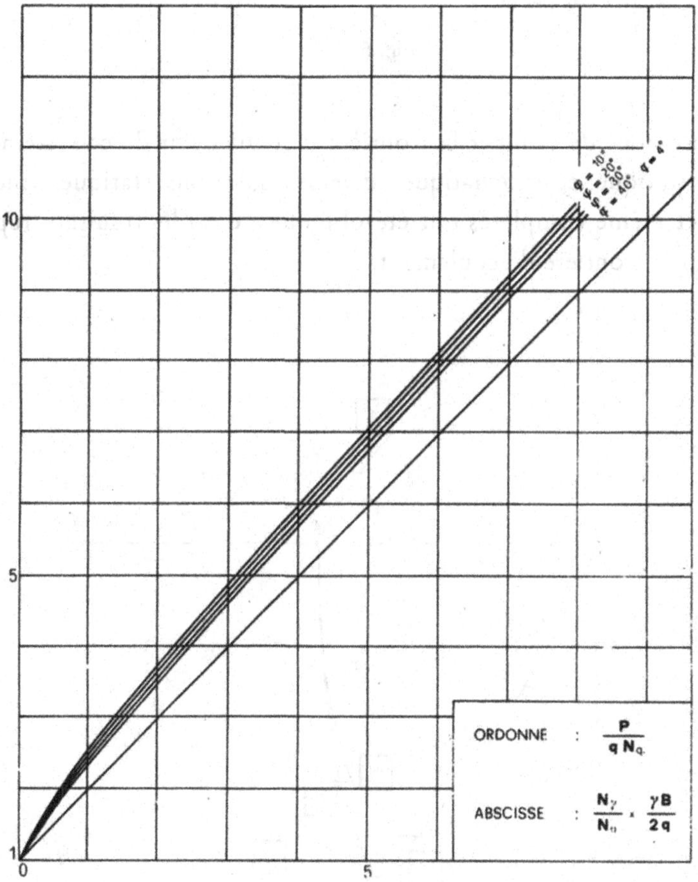

Fig. 4

2. *UN EXEMPLE D'ANALYSE LIMITE POUR LE MATERIAU STANDARD*

Le problème étudié est celui de la butée d'une paroi lisse sur un coin rectangulaire constitué d'un matériau de Tresca standard. Il y a *deux* paramètres de chargement indiqués sur la figure 5 ainsi que les paramètres de déformation associés.

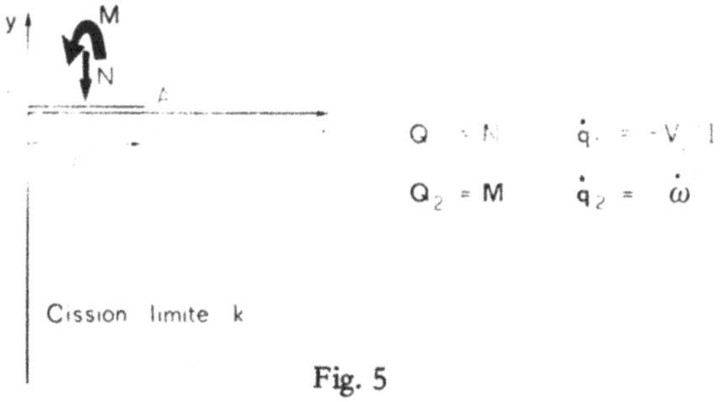

Fig. 5

On a utilisé pour déterminer la frontière d'écoulement de ce système standard les approches statique et cinématique: diverses solutions statiques, incomplètes, cinématiques, et même complètes ont été obtenues, d'où les résultats, représentés à la figure 6, pour la frontière d'écoulement.

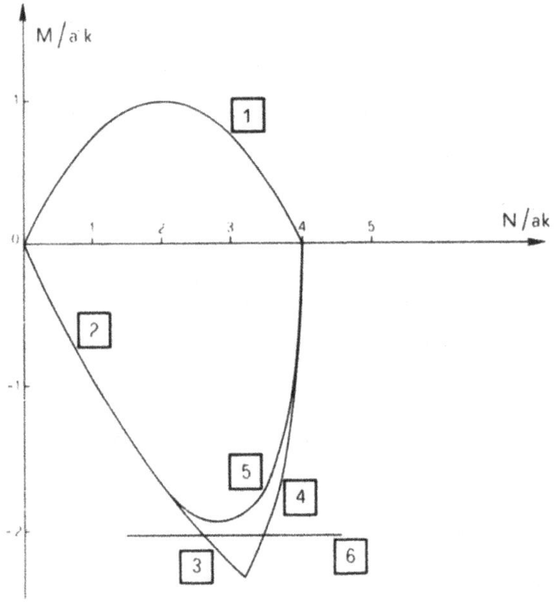

Fig. 6

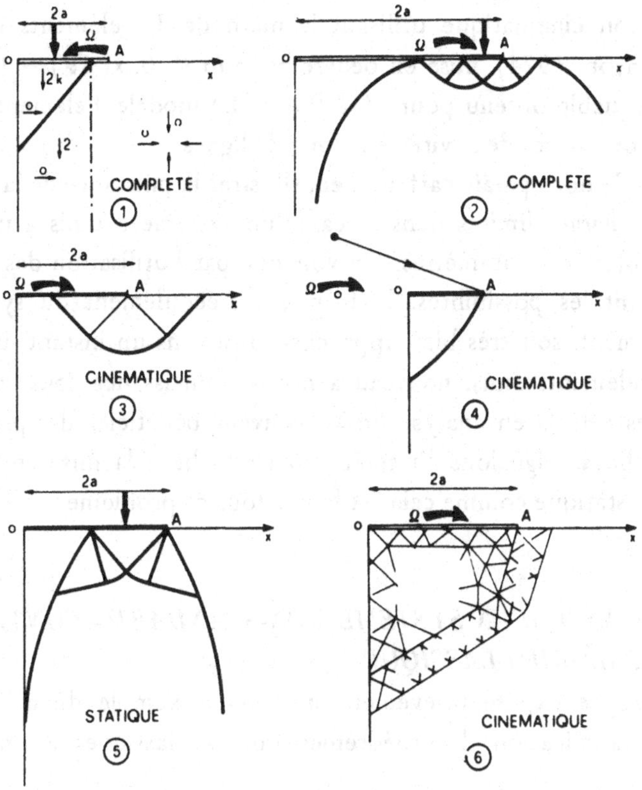

Fig. 7

La figure 7 schématise les solutions correspondant aux divers arcs obtenus.

(1): solution complète pour $\omega > 0$ et x(Ω) > 0.(force excentrée à gauche du milieu de la paroi). Il y a décollement sous ΩA.

(2): solution complète pour $\omega < 0$ et x(Ω) > 1.584 a (force excentrée à droite du milieu de la paroi: e > .792a). Il y a décollement sous Ω O.

(3): solution incomplète pour $\omega < 0$ et x(Ω) < 1.584a Il y a décollement sous Ω O si x(Ω) > 0.

(4): solution cinématique par cercles de glissement. Résultats numériques. L'approximation par l'extérieur obtenue ainsi est meilleure que celle de (3) pour 3.3 < N/ak < 4.0.

(5): solution statique utilisant les prolongements des champs de Prandtl par la méthode de Bishop. Résultats numériques. (Des explications détaillées concernant (1), (2), (3), (5) se trouvent dans les articles de l'auteur (1972, 1974)).

⑥: solution cinématique utilisant la méthode des éléments finis (Frémond, Pecker et Salençon, 1974), mise en oeuvre pour $\omega < 0$, $x(\Omega) = 0$ et $x(\Omega) = a$. Résultat remarquable obtenu pour $x(\Omega) = a$. Le modèle d'éléments finis employé admet des discontinuités de la vitesse le long de lignes.

Cet exemple nous paraît parfaitement illustrer la puissance de la formulation de la théorie des charges limites dans le cas d'un système soumis à un processus de chargement à plusieurs paramètres. On voit que par l'utilisation des divers types de solutions suivant les possibilités, la frontière d'écoulement du système est, soit connue exactement, soit très bien approchée. Insistons un instant sur l'emploi de la méthode des éléments finis, nouveau à notre connaissance sous cette forme, qui montre que les calculs en analyse limite peuvent bénéficier des progrès faits dans d'autres disciplines. Signalons l'intérêt qui s'attache à la mise en évidence d'une bonne solution statique comme cela est le cas pour ce problème.

3. *CHARGE LIMITE D'UN SYSTEME NON-STANDARD – CONTRE-EXEMPLE POUR LA THEORIE CLASSIQUE*

Ce paragraphe expose brièvement un contre-exemple dû à l'auteur (1972a, 1972b) dans l'application des théorèmes limites classiques à un système non--standard.

Le problème étudié est celui de la capacité portante d'une semelle filante peu profonde chargée axialement. Le sol est un matériau de Tresca *standard* (cission limite k) et la condition de frottement au contact entre sol et fondation est celle de Coulomb de coefficient $\mathrm{tg}\varphi > 0.39$. Le *système* est donc *non-standard.* Le matériau est supposé non pesant pour le calcul du terme de cohésion dans la capacité portante.

La figure 8 (droite) montre du champ de contrainte licite pour ce problème (la méthode de prolongement à l'infini utilisée est celle de Shield). La condition au contact sol-fondation est respectée. Ce champ correspond pour la capacité portante à:

$$F_s(h/a) = 2a\,k\,(\pi + 2 + h/a) \; .$$

La moitié gauche de la figure 8 présente un mécanisme de ruine dépendant du paramètre α. Pour un tel mécanisme on calcule sans ambiguïté la puissance

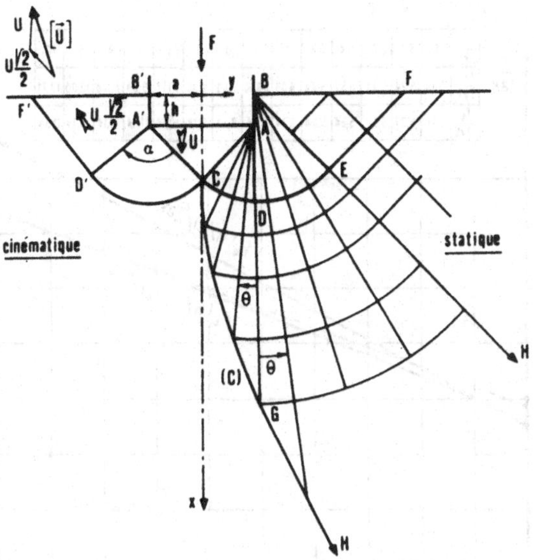

Fig. 8

dissipée:

en chaque point = puissance dans le champ de vitesses d'un tenseur des contraintes qui lui correspond par la règle d'écoulement:

- bien définie dans le matériau de Tresca standard;
- nulle sur AA' car pas de mouvement relatif;
- nulle sur A'B' car il y a décollement.

L'application du théorème cinématique classique ne présente donc pas de difficulté et conduit pour la force portante à:

$$F_c(h/a, \alpha) = 2ak(1 + 2\alpha + \mathrm{tg}\,(\frac{3\pi}{4} - \alpha) + \frac{h}{a}\frac{\sqrt{2}}{8}\frac{1}{\cos(3/4\pi - \alpha)},$$

dont le minimum en α pour chaque valeur de h/a est inférieur à F_s (h/a).

Ce paradoxe apparent, dû au caractère non-standard du système, montre bien qu'il y a lieu de prendre garde malgré leur aspect intuitif aux conditions d'applications des théorèmes du calcul à la rupture.

$\frac{h}{a}$	0	0,1	0,2	0,4	0,8	1	2	3	4	5	6	8	9	10
$\frac{F_c}{2ak}$	5,14	5,24	5,33	5,52	5,88	6,05	6,866	7,63	8,38	9,117	9,84	11,28	12,00	12,71
$\frac{F_s}{2ak}$	5,14	5,24	5,34	5,54	5,94	6,14	7,14	8,14	9,14	10,14	11,14	13,14	14,14	15,14

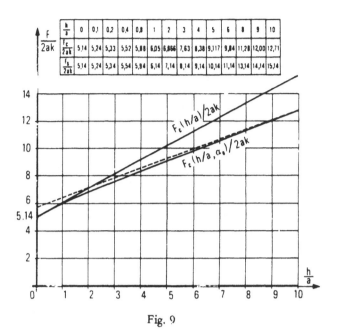

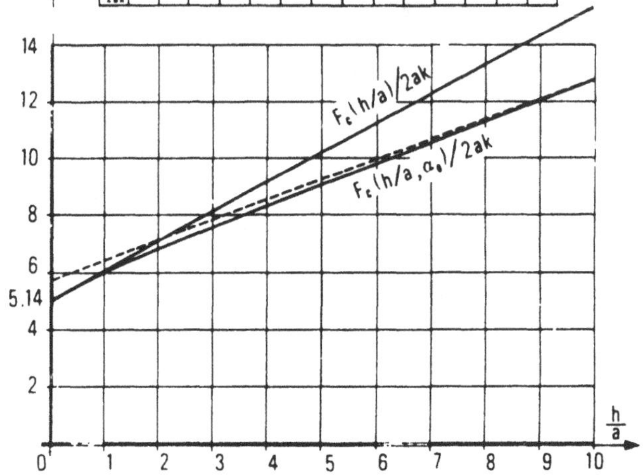

Fig. 9

4. CAPACITE PORTANTE D'UNE FONDATION SUR SOL NON-STANDARD

Ce paragraphe présente rapidement les résultats obtenus par Davis et Booker (1971) qui ont étudié la capacité portante d'une fondation superficielle sur un sol de Coulomb, standard ou non-standard en s'attachant à construire des solutions analogues aux solutions incomplètes pour le matériau standard.

Le sol homogène est défini par un critère de plasticité de Coulomb α, et un angle de dilatation correspondant à un potentiel plastique de Coulomb, ν.

Pour la construction de la solution on utilise les propriétés indiquées au chapitre VI à propos des lignes caractéristiques pour les contraintes et des lignes d'extension nulle. Cette construction est néanmoins plus délicate que dans le cas du matériau standard puisqu'il y a lieu de travailler au début avec les lignes caractéristiques des containtes, puis avec les caractéristiques des vitesses. Davis et Booker ayant construit la solution globale –couplage de γ et de q – pour diverses valeurs de φ, affirment que la condition de positivité de λ dans la loi de comportement est toujours vérifiée. Ce point vaudrait d'être examiné de plus près car il est contredit par le travail de Drescher (1971, 1972) dans le cas particulier $\gamma = 0$.

Du point de vue numérique, Davis et Booker ont trouvé que la capacité portante ainsi calculée variait peu quand ν variait entre 0 et φ. Ainsi le coefficient N_γ prend les valeurs suivantes:

$$\varphi = 30^\circ : \quad \nu = 30^\circ \quad N_\gamma = 14.96$$
$$\nu = 20^\circ \quad N_\gamma = 14.94$$
$$\nu = 10^\circ \quad N_\gamma = 14.92$$
$$\nu = 0^\circ \quad N_\gamma = 14.80 .$$

BIBLIOGRAPHIE

E.H. DAVIS & J.R. BOOKER (1972), The bearing capacity of strip footings from the standpoint of plasticity theory. Research report, Civ. Eng. Lab. Univ. Sydney.

A. DRESCHER (1971), A note on plane flow of granular media. Problèmes de la Rhéologie, Symp. Franco-Polonais, Varsovie 1971, p. 135-144.

A. DRESCHER (1972), Some remarks on plane flow of granular media. Arch. of Mech., vol. 24, n. 5-6, p. 837-848.

P. FLORENTIN et Y. GABRIEL (1974), Force portante d'une fondation sur sol verticalement non homogène. Travail fin d'études E.N.P.C., Laboroire de Mécanique des Solides, Ecole Polytechnique, Paris.

M. FREMOND, A. PECKER et J. SALENÇON (1974), Méthode variationnelle pour le matériau rigide plastique. Symposium Franco-Polonais de Rhéologie et de Mécanique des Sols, Nice, juillet 1974.

J. SALENÇON (1972), Butée d'une paroi lisse sur un massif plastique: solutions statiques. J. Mécanique, vol. 11, n. 1, p. 135-146.

J. SALENÇON (1972a), Un exemple de non validité de la théorie classique des charges limites pour un système non-standard. Discussion Int. Symp. on Foundations of Plasticity, Varsovie 1972, North Holland Pub. Cy.

J. SALENÇON (1972b), Charge limite d'un système non-standard. Un contre-exemple pour la théorie classique. Séminaire Plasticité et Viscoplasticité 1972, ed. D. Radenkovic et J. Salençon, Ediscience, Paris, 1974, p. 427-430. J. SALENCON (1974), Quelques résultats

J. SALENÇON (1974), Quelques résultats théoriques concernant la butée d'une paroi lisse sur un massif plastique. Annales ITBTP, n. 313, série TMC, n. 165, janvier 1974, p. 186-194.

J. SALENÇON, P. FLORENTIN et Y. GABRIEL (1976), Capacité portante globale d'une fondation sur un sol non-homogène, géotechnique, vol. 26, n. 2, p. 351-370.

NON-LINEAR VISCOUS SOIL BEHAVIOUR

L. Šuklje
University of Ljubljana

Introduction

In classical soil mechanics the deformations of soil bodies are computed according to the theory of elasticity while their safety against failure is examined with the assumption that soils are either elasto-plastic or rigid-plastic bodies. Their viscous properties are disregarded.

The recent development of procedure for calculation such as the finite element method, largely made practicable by electronic computers, has provided means for solving soil mechanics problems on the base of complex rheological relationships corresponding to the real soil behaviour.

The improvement of laboratory testing technique and construction of new testing devices such as the "true" triaxial apparatus, facilitated studies of evaluating and representing the stress-strain characteristics of earth materials.

For non-viscous soils, several non-linear rheological models have been elaborated in the past decade. They are of either hypoelastic or hyperelastic or elasto-plastic type. The application of available numerical procedures requires step by step linearization of non-linear problems. Thus, the constitutive laws have to be expressed in incremental form.

Since the higher grade hypoelastic models are relatively complex for incorporation, the so called "hybrid hypoelastic models" (Desai, 1972) have been extensively used. They represent an incremental form of linear elastic constitutive laws; the elastic matrix relating increments of stress and strain tensors, contains stress and/or strain dependent tangent values of elastic parameters as obtained by differentiating the stress-strain relationships corresponding to experimental data. Several comparisons between the observed field behaviour and calculated results were made by using the "elastic model" as elaborated at the University of California, Berkeley (Kulhawy, Duncan and Seed, 1969; Duncan and Chang, 1970). This model is based on Kondner's (1963) hyperbolic stress-strain relationship, on Janbu's (1963) parabolic expression for the initial values of tangent moduli in terms of hydrostatic pressures, and on a hyperbolic lateral versus axial strain relationship. The tangent elastic parameters are expressed in terms of the lateral pressure σ_3 and the stress difference $\sigma_1 - \sigma_3$. The model, however, is restricted to work hardening soils and to conventional triaxial test conditions with lateral pressures σ_3 kept constant; furthermore, swelling and settlement phases of the volume variation cannot be distinguished and in the perfect plasticity the deformations do not occur at constant volume.

Darve (1974) has succeeded in elaborating a constitutive law of hyperelastic type valid in a wide range of stress paths and representing the progressive rupture of soils in a correct way. It describes continuously the behaviour of soils from the small quasi elastic deformations up to the domain of perfect plasticity, both for loading and unloading, and the eventual reduction of the strength from the peak to the residual value has been taken into account. However, owing to the great number of parameters to be determined experimentally, the model in its general form can hardly be applied in practical computations.

In a similar way, yet in terms of octahedral values an empirical stress-strain law has been proposed by Pregl (1974), while Saje (1974), Majes (1974) and Vidmar (1974) have presented experimental data corresponding to some selected stress-paths (Šuklje, 1967), in the form of steps of charts relating octahedral values of stresses and strains; the tangent values of elastic parameters have been deduced by applying spline functions.

The strain energy function may be expressed in terms of powers of strain. According to the principle of conservation of energy in an adiabatic process, stress components can be obtained as partial derivatives of the strain energy function with respect to the corresponding strain components. In this way, a second order stress-strain law of the hyperelastic type was deduced by Chang, Ko, Scott and Westmann (1968); they found, however, that this law only approximates the real material for a limited range of strain states. The linearized incremental version of the third-order strain-stress law as deduced by Masson (1971) by using a complementary energy density function, has been found to model the medium-loose sand with reasonable accuracy.

In the elastic-plastic analysis, the total strain increments are assumed to be divisible into elastic and plastic components. The increments of stress are assumed to be related to the increments of elastic strains by means of a symmetrical stress-dependent elasticity matrix, while the plastic strain increments may be expressed according to Drucker's (1951, 1964) "normality rule" by the partial derivatives of the plastic potential function with respect to the corresponding stress components. The yield surface defines the boundary between the state of stress where only elastic strain occurs (inside the yield surface) and the state of stress where both elastic and plastic strain occurs (outside the yield surface).

The classical formulation of theory of plasticity considers a class of materials for which the yield function also serves as the plastic potential for the flow. An example of such an elasto-plastic model with the "associative flow rule" is the adaption of the critical state soil mechanics theory (Roscoe, Schofield and Wroth,

1958, and further publications by Roscoe and his cooperators) for use in finite elements as developed by Zienkiewicz and Naylor (1971).

Now, for sandy soils and for normally consolidated clays the plastic potential surface was proved not to coincide with the yield surface. For work hardening materials Lade (1976, see also Lade 1972, Ozawa 1973) considered the strain increments to consist of an elastic component, a plastic collapse component derived from a cap shaped associative yield surface, and a plastic expansive component obeying a non-associative flow rule with curved cone shaped yield and plastic potential surface. The resulting matrix of incremental rheological relations contains 14 parameters; all of them can be derived from isotropic compression and conventional drained triaxial tests.

In none of the above mentioned rheological models the viscous soil properties were considered. So far, the viscous soil behaviour was investigated mainly in linear strain conditions of oedometer tests, and, to a lesser extent, in rotationally symmetric stress conditions of conventional triaxial tests. For arbitrary states of stress, simple (mainly linear) models containing few rheological elements connected either parallelly or/and in series, have been proposed more or less by intuition. A critical review of all three groups of rheological relationships for soils exhibiting viscous properties is presented in the first part (I) of these lectures.

When analyzing the development of effective stresses and displacements in poorly permeable soils, one has to consider the equation of the transient fluid flow, the effective stress principle, the rheological relationships of soils, the equilibrium equations, and the relations between displacements and strains. For elastic saturated media the above relations lead to the well known Biot equation (1935 and 1941). The solutions of this equation are available for circular, square, strip and rectangular footings (e.g. solutions by Josselin de Jong 1957, Mandel 1957 and 1961. McNamee and Gibson 1960, Schiffman et all. 1969, Gibson et all. 1970, etc). For complex boundary conditions the finite element formulation based on the variational principle has been proposed by Sandhu and Wilson (1969) and the procedure extended to include a logarithmic time increment by Hwang et all. (1971). Numerical solutions of Biot's equation of consolidation have also been given for some simple rheological models (e.g. by Kisiel and Lysik, 1966, for the M/V model and plane state of strain, and by Zaretsky, 1967, for the Kelvin body and a surface point load). For soils represented by a linear rheological model consisting of the Hookean spring connected in series with the Kelvin body, Booker and Small (1977) presented recently a numerical solution using the method of finite elements.

For soils exhibiting non-linear viscous properties, the joint solution of the diffusion and equilibrium equations is still in a formulative state of development. In the present lectures, only separate analyses of the total stress-strain state and of the consolidation process are given. The second part (II') of the lectures contains first a numerical procedure enabling the computation of stresses and displacements as developing during loading and afterwards, provided that the seepage resistance is neglected. The following consolidation analyses, introduced by the deduction of a simplified form of the diffusion equation for soils, concern (a) a one-dimensional study of the influence of factors governing the consolidation of non-linear viscous soils, (b) two cases of the approximate two-dimensional computation of the development of displacements in the half-plane.

I. VISCOUS RHEOLOGICAL MODELS

A. Viscous Rheological Models in Linear State of Strain

1. RELATIONSHIPS OF THE TYPE $R(e, \dot{e}, \sigma') = 0$

Definition of Effective Stress for Viscous Soils

For viscous soils, the effective stresses have to be defined as that part of total stresses which governs the strains and their speeds. The difference between total and effective stresses, both related to a continuous medium, defines the neutral stress. If the hydraulic field is assumed to be a continuous potential field, the neutral stress in a saturated soil can be taken equal to the pore pressure provided that the seepage forces are considered as volume forces affecting the grain skeleton.

In a partly saturated soil the neutral pressure can be taken equal to

$$u = \chi\, u_w + (1 - \chi)\, u_a \tag{1.1}$$

u_w and u_a being the pressure of the fluid in the voids filled with water and air respectively, and $\chi \leqslant 1$ a coefficient to be determined experimentally (Bishop, 1955).

The Classical Oedometer Test

Classical methods of the computation of the consolidation of soil layers due to the effect of additional loading are based on the assumption that the stress states in soils correspond to linear strain states and that the strains depend only on normal stresses acting in the main strain direction. The majority of the consolidation analyses accounting for viscous soil properties have been related to the same assumption. Such rheological relationships can be obtained by the oedometer test as introduced by Terzaghi (1923).

According to the classical procedure in oedometer testing the laterally confined saturated cylindrical sample is subjected to successive normal loads σ_z in such steps that $\sigma_{z\,i+1} = 2\sigma_{z\,i}$. If the viscous effects are negligibly small the transition of the initial void ratio e_i at the beginning of the load step $(i + 1)$ into the void ratio e_{i+1} at the end of the step is determined by the horizontal asymptote of the consolidation curve $e = e(\log t/t_o)$ (Fig. 1.1). At the end of each step the effective pressure σ' is approximately equal to the total pressure because the remaining pore pressures are immeasurably small. Terzaghi has ascertained that the relationship $e = e(\sigma')$ can be approximated by a half-logarithmic straight line (Fig. 1.2) provided the clay is normally consolidated:

$$e = e_\alpha - \alpha_e \log_{10} \frac{\sigma'}{\sigma'_\alpha} \tag{1.2}$$

or with

$$-\frac{\alpha_e}{2.3} = B, \qquad e_\alpha - \frac{\alpha_e}{2.3} \ln \frac{\sigma_o}{\sigma'_\alpha} = A,$$

$$e = A + B \ln \frac{\sigma'}{\sigma_o} \tag{1.2-a}$$

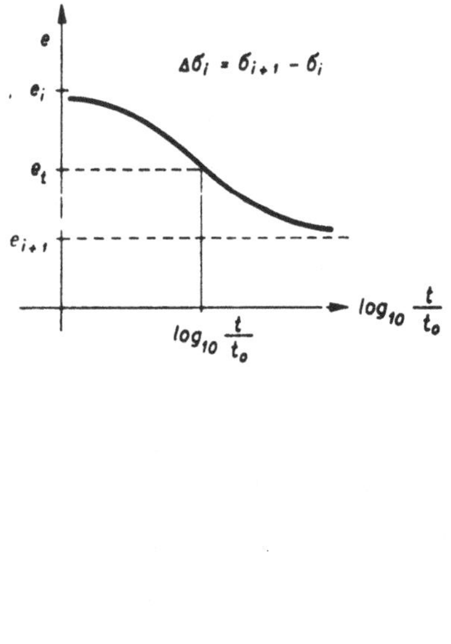

Fig. 1.1 - Consolidation line of a non-viscous sample for a suddenly applied load interval

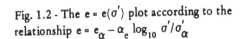

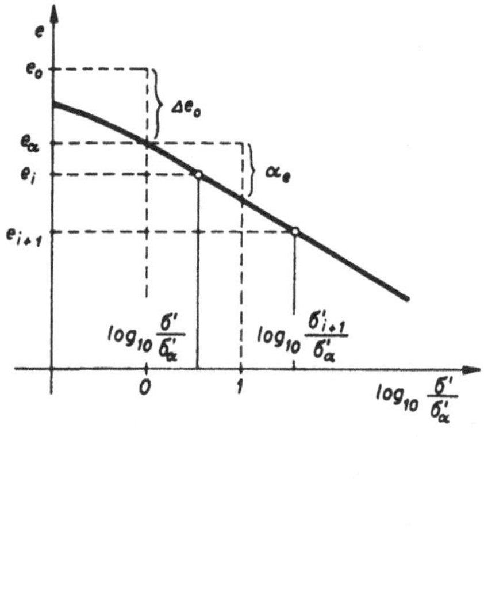

Fig. 1.2 - The $e = e(\sigma')$ plot according to the relationship $e = e_\alpha - \alpha_e \log_{10} \sigma'/\sigma'_\alpha$

Terzaghi himself observed that in the above defined classical test the asymptote of the curve $e = e(\log t/t_o)$ usually was not horizontal and that the primary consolidation during which the pore pressures had dissipated to immeasurably small values, was followed by a secondary consolidation of either great or small intensity and duration. Buisman (1936) found that in the coordinate system $(e, \log t/t_o)$ the secondary consolidation line was straight (Fig. 1.3).

$$e = e_\beta - \beta_e \log_{10} t/t_\beta \qquad (1.3)$$

or, with

$$\beta = \frac{\beta_e}{2.3}$$

$$e = e_\beta - \beta \ln t/t_\beta \qquad (1.3\text{-a})$$

The coefficient β depends on the effective pressure σ' :

$$\beta = \beta(\sigma') \qquad (1.4)$$

According to the experiments, the above relation can often be approximated by the equation

$$\beta = C + D \ln \frac{\sigma'}{\sigma_o} \qquad (1.4\text{-a})$$

The speed of the void-ratio change in the secondary consolidation phase can be obtained by the differentiation of Eq. (1.2–a).

$$\frac{\partial e}{\partial t} = \dot{e} = -\frac{\beta}{t} \qquad (1.5)$$

Thus Eq. (1.3-a) may be written also in the form

$$e = e_\beta + \beta \ln \frac{\dot{e}}{\dot{e}_\beta} \qquad (1.6)$$

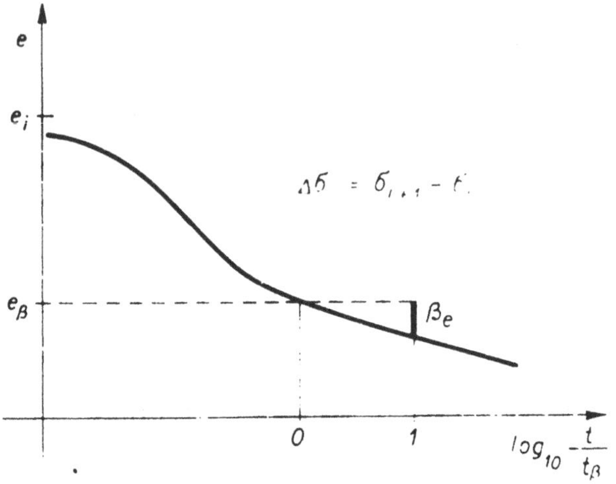

Fig. 1.3 - Consolidation line according to Buisman (1936)

Isotaches of the Oedometer Test

(a)

When in single load steps the consolidation curves have not horizontal asymptotes, the plots $e = e(\sigma')$ are not unambiguously defined. However, their meaning becomes clear if they are related to certain arbitrarily chosen consolidation speeds $\dot{e}_i = 10^{-i}\ \dot{e}_\beta$, i = 0, 1, 2, . . ., or to the respective time (according to Eq. 1.5):

$$t_i = -\frac{\beta}{\dot{e}_i} \tag{1.7}$$

(At the volume decrease the speeds $\dot{e}_i$ are negative.) Such plots

$$e = [e(\sigma')]_{\dot{e} = \text{const}} \tag{1.8}$$

may be called isotaches of the void ratio change or, briefly, isotaches (i.e. curves of equal speed).

When in the coordinate system $(e, \log \sigma'/\sigma'_\alpha)$ the isotache $\dot{e} = \dot{e}_\beta$ is a straight line and the secondary consolidation obeys Buisman's law, the equation of isotache sets can be obtained by combining Eqs (1.2) and (1.6):

$$e - \beta \ln \frac{\dot{e}}{\dot{e}_\beta} = A + B \ln \frac{\sigma'}{\sigma_o}$$

yielding, with $\dot{e}_\beta \equiv \dot{e}_o$,

$$e = A + B \ln \frac{\sigma'}{\sigma_o} + \beta \ln \frac{\dot{e}}{\dot{e}_o} \tag{1.9}$$

or

$$\dot{e} = \dot{e}_o \exp \frac{A + B \ln \frac{\sigma'}{\sigma_o} - e}{\beta} \tag{1.10}$$

If β is governed by Eq. (1.4-a), then

$$\dot{e} = \dot{e}_o \exp \frac{A + B \ln \frac{\sigma'}{\sigma_o} - e}{C + D \ln \frac{\sigma'}{\sigma_o}} \tag{1.11}$$

$\dot{e}_o$ and σ_o are constants arbitrarily chosen in the domain of values observed during the secondary consolidation of oedometer samples, while A, B, C and D are parameters to be determined from four groups of data for e, $\dot{e}$ and σ' in the domain of the observed secondary consolidation lines.

At a sudden step-load increase negative thixotropic effects can appear and cause the failure of the structure of the grain skeleton and a change in the deformability of the soil. In order to avoid such failure, the classical way of the step loading is recommended to be replaced by a slow, continuous load increase, interrupted by creep observation of long duration at certain values of effective stresses σ'. If in the secondary consolidation phase the void ratios in this test procedure are also observed to decrease with the logarithm of time, and if the corresponding isotaches are again straight in the coordinate system ($\log \sigma'/\sigma_o$, e), the expressions (1.9) and (1.10) remain valid.

In the primary consolidation phase the plot e = e(t) represents the development of the mean values of the void ratios e with time; these void ratios correspond to certain mean values of effective stresses σ' :

$$(1.12) \qquad \bar{\dot{e}} = \bar{\dot{e}}(t) \ , \qquad \bar{\dot{e}} = \bar{\dot{e}}(\bar{\sigma}')$$

If we assume a certain form of isochrones and consider Darcy's seepage law $v = k\,i$ (v = filtration speed, k = coefficient permeability, i = hydraulic gradient), the expression for the mean value of the excess pore-pressure, u, can be obtained by equalizing the speed of the volume change and the output seepage speed of the pore water at the drained boundary of the sample (see Šuklje, 1957 and 1969-a).

Now, the isotaches for speeds occurring during the primary consolidation of thin oedometer samples are hardly of any interest for the computation of the consolidation of natural layers whose thickness is by far greater. Even the speeds at the very beginning of the primary consolidation of thicker layers are usually smaller than the speeds observed in the early phase of the secondary consolidation of oedometer samples.

The equality of the output seepage speed of the pore-water and the volume change speed governs, of course, also the secondary consolidation of saturated samples. Now, in this phase of consolidation the excess pore-pressures conditioning the seepage of the pore water are so small that the effective pressures can, with sufficient accuracy, be taken equal to the total pressures. Thus, the computation of pore pressures can be omitted.

If we can extrapolate the line of the observed secondary consolidation beyond the time of observation with an acceptable probability, the coordinated values of effective pressures, void ratios and their speeds can be obtained for a sufficiently long period of consolidation. Several groups of such coordinated values permit the construction of isotaches. Corresponding to their form they can be represented by appropriate analytical functions.

The coefficient β appearing in Eq (1.9) and (1.10) determines, at a given value of σ', the uniform distance of isotaches corresponding to successive speeds $e_i = 10^{-i}\,\dot{e}_\beta$, $i = 0, 1, 2, \ldots$ (Fig. 1.4). In the coordinate system ($\ln \sigma'/\sigma_0$, e) the isotaches are straight lines if, according to Eq. 1.4-a), β is a linear function of $\ln \sigma'/\sigma_0$. When β is constant (i.e. D in Eq. 1.4-a equal to zero), such isotaches become parallel lines. If β is a non-linear function of $\ln \sigma'/\sigma_0$, the isotaches get curved in the ($\ln \sigma'/\sigma_0$, e) coordinate system except the isotache $\dot{e} = \dot{e}_0$ which remains a straight line.

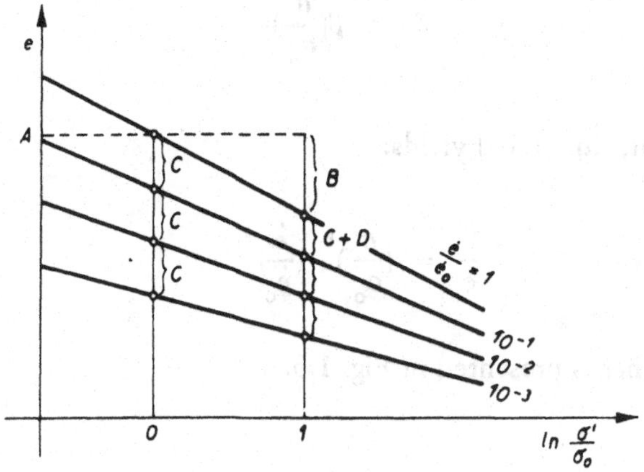

Fig. 1.4 - Isotaches according to Eq. (1.11)

(b)

For $\sigma' = 0$, Eq. (1.9) is indefinite. In order to obtain definite values of void ratios also in this case, the equation of isotaches can alternatively be given the form:

$$e = A + B \ln \frac{\sigma' + \sigma_o}{\sigma_o} + \beta \ln \frac{\dot{e}}{\dot{e}_o} \tag{1.13}$$

(c)

The validity of Eq. (1.9) is restricted also by the limitation in the decrease of void ratios which cannot get negative values. If we replace Buisman's law (Eq. 1.3) by the assumption that during the secondary consolidation the logarithm of void ratios decrease linearly with the logarithm of time (Bent Hansen, 1969), similar deductions as presented under (a) lead to an analogous expression for isotaches:

$$\frac{\dot{e}}{\dot{e}_o} = \exp \frac{A + B \ln \frac{\sigma'}{\sigma_o} - \ln e}{\beta} \tag{1.14}$$

or, alternatively,

$$\frac{\dot{e}}{\dot{e}_o} = \exp \frac{A + B \ln \frac{\sigma' - \sigma_o}{\sigma_o} - \ln e}{\beta} \tag{1.15}$$

$$\beta = \beta(\frac{\sigma'}{\sigma_o})$$

When β = constant, Eq. (1.14) yields:

$$\frac{e}{e_o} = (\frac{\sigma'}{\sigma_o})^b \, (\frac{\dot{e}}{\dot{e}_o})^c \tag{1.16}$$

Such a set of isotaches is presented in Fig. 1.5.

(d)

The experimental results can often satisfactorily be expressed by the series:

$$e = \sum_{i=0}^{n} [a_i(\frac{\sigma'}{\sigma_o})^i] - [\ln \frac{\dot{e}}{\dot{e}_o}]^{1/c} \sum_{i=0}^{m} [b_i\,(\frac{\sigma'}{\sigma_o})^i] \tag{1.17}$$

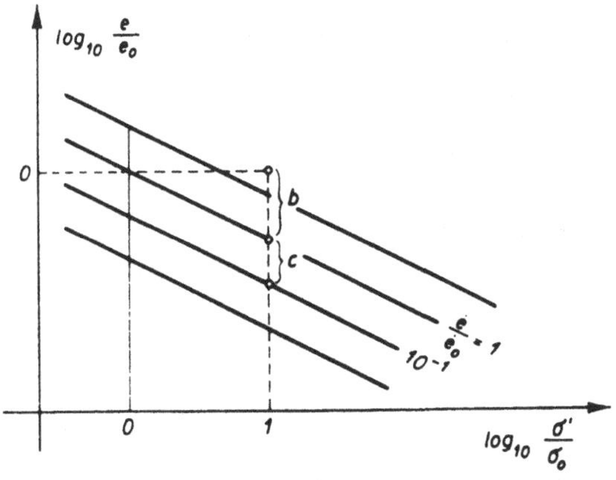

Fig. 1.5 - Isotaches according to Eq. (1.16)

In order to determine n parameters a_i, m parameters b_i and the parameter c, Eq. (1.17) has to be applied for (m + n +1) groups of data for the corresponding values of e, $\dot{e}$ and σ'. These data can be obtained either immediately from the secondary consolidation curves of the oedometer test or from the isotaches if they had previously been plotted by using the data of consolidation curves.

When $c = 1$, the secondary consolidation lines are straight in the $(\log t/t_o, e)$ coordinate system. When $c > 1$, porosity decreases slower than with the logarithm of time at approximately constant effective pressure σ' of the secondary consolidation.

Eq. (1.17) as well as other analytical expressions for isotache sets can be applied to single sections of the entire isotache set by considering the continuity conditions in the jointing points, requiring equal values for void ratios e and for their speeds $\dot{e}$.

(e)

Poorooshasb (1969) and Sivapatham have analysed the consolidation of saturated viscous soils by assuming that their rheological properties obey the equation

$$e = e_o - A(\sigma' - \sigma_o) + B\left(\frac{\dot{e}}{\dot{e}_o}\right)^a \left(\frac{\sigma' - \sigma_o}{\sigma_o}\right)^b \tag{1.18}$$

Another example of the isotache equation is the expression used by Hawley and Borin (1973) in their "unified theory of consolidation":

$$\frac{\partial e}{\partial t} = B\left[1 - \frac{e_L - e_u}{e_L - e}\right]$$

where

$$e_L = \Gamma_L - \lambda_L \log \sigma'$$

$$e_u = \Gamma_u - \lambda_u \log \sigma'$$

As the parameters $\Gamma_L, \Gamma_u, \lambda_L$ and λ_u are constant values, the above equations can be written in the form:

$$\dot{e} = A - \frac{C - D \ln \dfrac{\sigma'}{\sigma_o}}{E - F \ln \dfrac{\sigma'}{\sigma_o} - e} \tag{1.19}$$

Kelvin's Rheological Model

Several consolidation studies for soils exhibiting viscous properties are based on rheological relationships which correspond to the Kelvin rheological model (Fig. 1.7). So the linear Kelvin body represents the model of the relationship used in the

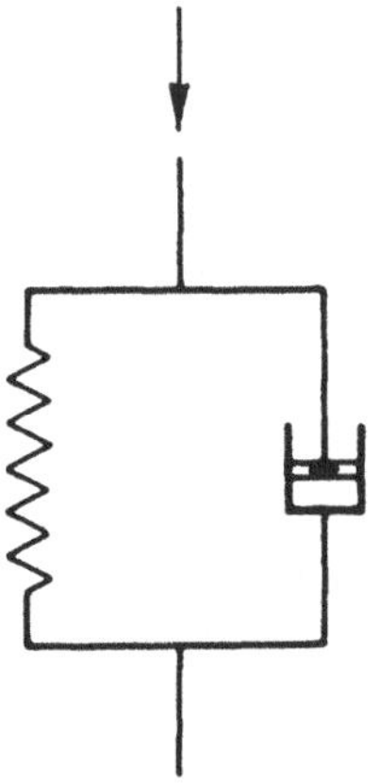

Fig. 1.6 - Kelvin's rheological model

first complete consolidation analysis accounting for viscous effects, i.e. in the so called Theory B of Taylor (1942). Stroganov (1963) uses the Kelvin model with the non-linear Hooke and with the linear Newton element, and Barden (1965) with the linear Hooke and non-linear Newton element.

The speed of the effective stress does not intervene in the Kelvin rheological equation and does not appear in any rheological model whose elements are connected in parallel. Thus, the rheological equation of Kelvin's body is of the type $R(e, \dot{e}, \sigma') = 0$

A wide range of the observed secondary consolidation of oedometer samples and of possible extrapolations of observed secondary consolidation lines can be covered by considering the soil as a non-linear Kelvin body according to the equation

$$e = e_0(\sigma') - A(\sigma') \left[1 - (\dot{e}/\dot{e}_0)^n\right] \tag{1.20}$$

In special cases the reference curve $e_0 = [e_0(\sigma')]_{\dot{e}=\dot{e}_0}$ and the parameter $A = A(\sigma')$ can be expressed by analytical functions. In order to meet any possible variety of experimental data, they can be presented by sets of point values; the interpolation can be made by using spline functions (Desai, 1971). The distance between two subsequent isotaches

$$\Delta e_i = (e)_{\dot{e}/\dot{e}_0 = 10^{-i}} - (e)_{\dot{e}/\dot{e}_0 = 10^{-(i+1)}}, \qquad i = 0,1,2,\ldots$$

is

$$\Delta e_i = A\, 10^{-in}\,(1 - 10^{-n}) \tag{1.21}$$

With

$$\Delta e_{i=0} = \Delta e_0 = A\,(1 - 10^{-n})$$

denoting the distance between the isotaches corresponding to $i = 0$ and $i = 1$, Eq. (1.21) yields:

$$\Delta e_i/\Delta e_0 = 10^{-in}, \qquad \Delta e_{i=0} = 10^{-i}, \qquad i = 0,1,2,\ldots \tag{1.22}$$

With increasing parameter i the ratio $\Delta e_i/\Delta e_0$ increases when $n < 0$, and decreases when $0 < n < 1$. For different values of n the ratios $\eta = \Delta e_i/\Delta e_0 = \eta(i)$ are presented, for the interval $0 \leqslant i \leqslant 8$, in Fig. 1.7.

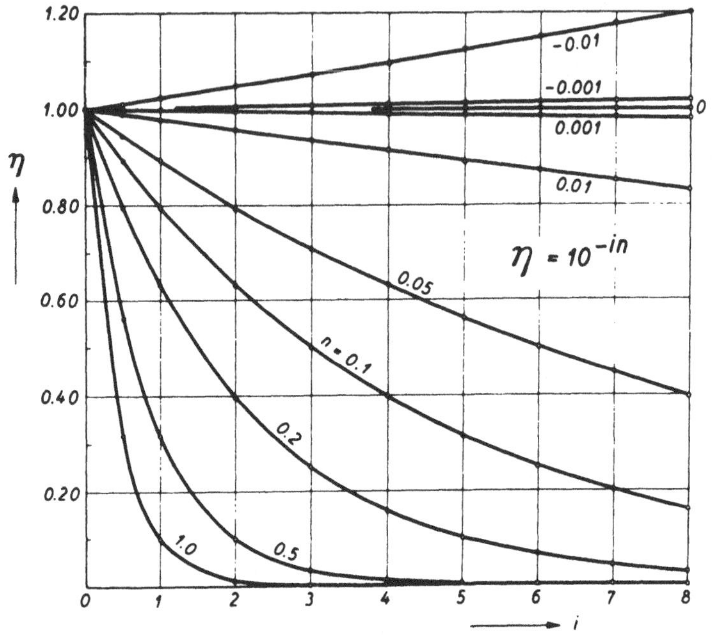

Fig. 1.7 - Plots $\eta = \Delta e_i/\Delta e_0 = \eta(i, n)$ according to Eq. (1.22)

With $e_0 = e_0(\sigma')$ being a linear function ($e_0 = a + b\,\sigma'$) and $A = A(\sigma') = \text{const}$, Eq. (1.20) represents Barden's form of the non-linear Kelvin body including, at $n = 1$, the linear Kelvin body of Taylor's Theory B. The isotaches become parallel straight lines (Fig. 1.8), the distances between the adjacent isotaches $\dot{e}_i = 10^{-i}\,\dot{e}_0$, $i = 0,1,2,..$ being governed by the parameter n. The asymptotic isotache has the equation:

$$e = (a - A) + b\,\sigma' \tag{1.23}$$

With $A = e_0$, Eq. (1.20) reduces to

$$e = e_0(\dot{e}/\dot{e}_0)^n\,, \quad n = \frac{\beta}{1+\beta} \tag{1.24}$$

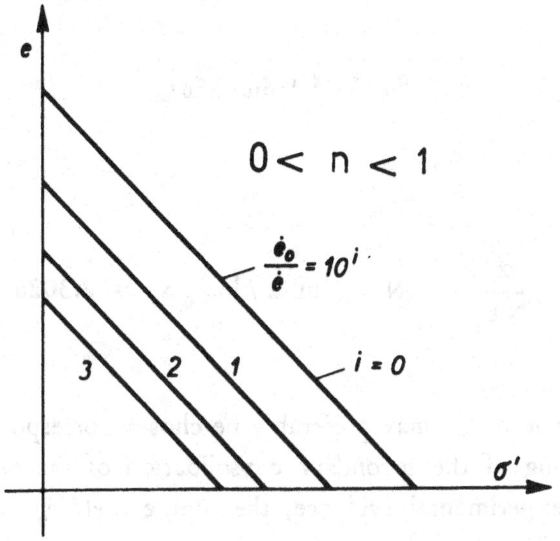

Fig. 1.8 - Isotaches according to Barden's rheological model

corresponding to the power law of the secondary consolidation

$$e = e_0(t_0/t)^{\mu} \quad \text{or} \quad \log_{10} e = \log_{10} e_0 - \beta \log_{10}(t/t_0)$$

With

$$A(\sigma') = B\,[(\sigma' - \sigma_0)/\sigma_0]^b \,, \qquad e_0(\sigma') - A(\sigma') = e_0 - A\,(\sigma' - \sigma_0)$$

and n ≡ a, Eq. (1.20) covers the rheological relationship of Poorooshasb and Sivapathan (Eq. 1.18).

The equidistant isotaches at $n \to 0$ (see Eq. 12) correspond to Buisman's law of the secondary consolidation

$$e = e_0 - \alpha \log_{10}(t/t_0) \tag{1.26}$$

yielding

$$(1.27) \qquad e = e_0 + \alpha \log_{10}(\dot{e}/\dot{e}_0)$$

if the relations

$$(1.28) \qquad \dot{e} = -\frac{\alpha}{N\,t} \;, \quad N = (\ln x / \log_{10} x) = 2.3026$$

are considered.

The reference speed $\dot{e}_0$ may preferably be chosen corresponding to a time t_0 at the very beginning of the secondary consolidation of the oedometer sample. According to the experimental evidence, the plot $e = e(\dot{e}/\dot{e}_0)$ according to Eq. (1.20) can be considered to have, at $\dot{e} = \dot{e}_0$, common tangent with the plot $e = e(\dot{e}/\dot{e}_0)$ according to Eq. (1.27). Consequently:

$$(1.29) \qquad \frac{de}{d\eta} = A\,n\,(\eta)^{n-1} = \frac{\alpha}{N}(\eta)^{-1} \;, \qquad \eta = \dot{e}/\dot{e}_0$$

yielding, at $\dot{e} = \dot{e}_0$,

$$(1.30) \qquad A = \frac{\alpha}{n\,N}$$

According to Eq. (1.27), α denotes the decrease in void ratio along the line (1.27) for the difference in $\eta = \dot{e}/\dot{e}_0$ from 10^0 to 10^{-1}. The parameter n governs the extrapolation of the secondary consolidation lines beyond the time of observation. Choosing the right parameter is a delicate task which is elleviated if the history of geological loading is known, and the void ratio of the sample as well as of samples of the same soil from greater depths has carefully been determined (example in Bjerrum, 1967).

From Fig. 1.7 it is evident that for $n = 1$ the isotaches $i > 3$ practically condense to a single line (the differences between subsequent isotaches approach zero value); then viscous soil properties have no remarkable influence onto the consolidation process. On the other hand, at $-0.001 < n < 0.001$ the subsequent isotaches i = 0,1,2,.. become practically equidistant indicating the occurrence of the secondary consolidation according to Buisman's logarithmic law of the secondary consolidation.

Linear isotaches according to Fig. 1.8 can be applied only as a rough approximation for a sufficiently small stress interval. By considering the non-linearity of the Hooke element, Stroganov (1963) tried to enlarge the stress domain of the validity of solutions of the consolidation process. However, since the Newton element remained linear, viscous effects cannot influence the consolidation of thicker layers.

De Josselin de Jong (1968) represents viscous grain skeleton by a generalized Kelvin model. The distribution of element parameters is continuous and stochastic. This means that the properties of individual elements vary at random and that their abundance enables representing the frequency of occurrence of their parameters by a continuous function. The settlement-time curves have been approximated alternatively by the combination of several straight lines if plotted in a double logarithmic diagram, i.e. according to the "sectional power time law", or according to Buisman's secular effect.

2. RELATIONSHIPS OF THE TYPE $R(e, \dot{e}, \sigma', \dot{\sigma}') = 0$

Rheological Models Connected in Series

The first consolidation theory accounting for secondary (viscous) effects, is to be attributed to Taylor and Merchant (Theory A, 1938). Basic assumptions of this theory correspond to rheological properties of a model obtained by the connection in series of the Hookean spring with Kelvin's body (Fig. 2.1). Later, this theory was further developed by Gibson and Lo (1961) as well as by Florin (1961). The fundamental rheological relationship of such a model applied to the conditions of the oedometer test is:

$$e = e_o + A\dot{e} + B\sigma' + C\dot{\sigma}' \tag{2.1}$$

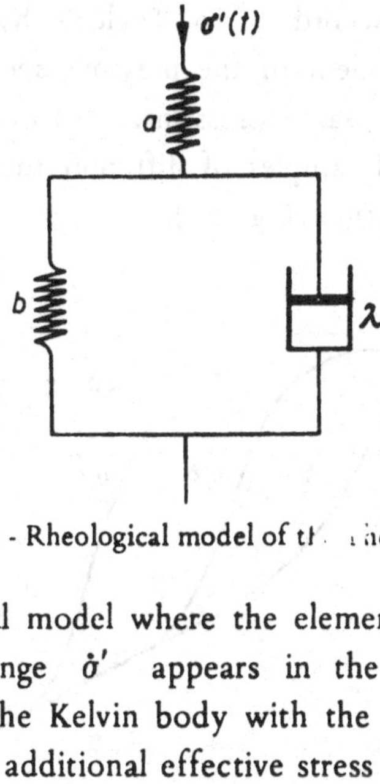

Fig. 2.1 - Rheological model of tl. ı ıeory A

As in any rheological model where the elements are connected in series, the speed of the stress change $\dot{\sigma}'$ appears in the rheological relationship. The connection in series of the Kelvin body with the Hookean spring forms a body which cannot assume any additional effective stress without additional strain, while in a pure Kelvin body the effective stress can increase immediately after a sudden load application. The investigators applying the above connection in series seem to have wanted to avoid this property of the pure Kelvin body which could not be accorded with the classical definition of the principle of effective stress (see e.g. the interpretation of Berre and Iversen, 1972). Now, the classical definition did not apply for viscous soil properties. Naturally, the model of the Theory A cannot express the effect of the "viscous resistance" (Taylor, 1942) allowing, at zero initial displacement, the effective stress increase proportional to the initial strain speed. This effect of the viscous resistance has been proved by laboratory (Šuklje, 1957) as well as by field (Bjerrum, 1967) observations.

By the above treated connection in series, the limited possibility of the linear Kelvin model to express viscous effects on the consolidation of thicker layers, has not been improved (Šuklje, 1969-a).

Rheological Relationships Obtained by Differentiating the Equation of Time Lines

Taylor (1942) suggested representing viscous effects on the deformability of oedometer samples by "time lines", i.e. void ratio e versus effective pressure σ' plots corresponding to certain times t_i passed after a sudden load increase. Times t_i have to be chosen in the phase of secondary consolidation which occurs at very small excess pore-pressures. According to Taylor's hypothesis, the time lines $e = [e(\sigma')]_{t_i = const}$ are independent of the previous speed of the effective stress increase. This means that at the same stress interval from σ_α to σ_β the primary consolidation curves $e = e(t)$ of samples of different thickness pass over into the same line of secondary consolidation (Fig. 2.2).

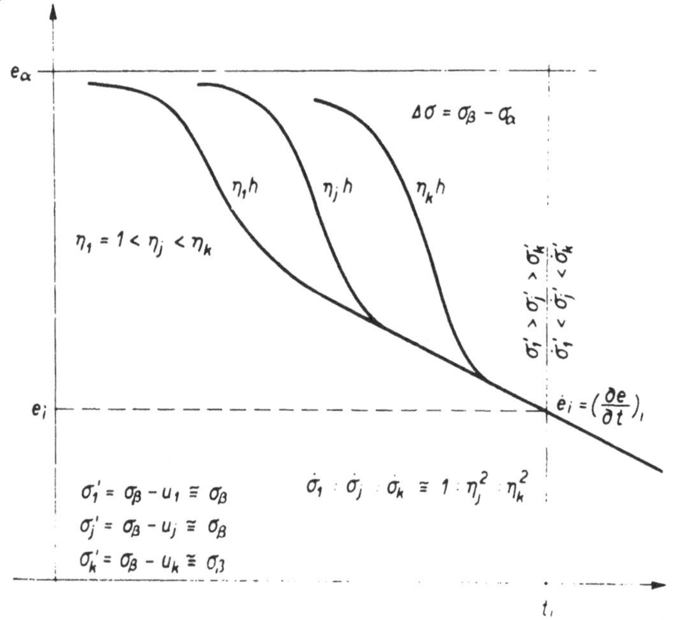

Fig. 2.2 - Consolidation curves of samples of different thickness for soils whose rheological relationships do not depend on the effective stress speed (Taylor's hypothesis)

Along this common secondary consolidation line the pore pressures of different samples differ approximately in proportions of squares of their thickness. However, all of them are very small and, consequently, the consolidation of different samples occurs, in this phase, at practically the same effective stress σ'. At any time of this consolidation phase, the void ratio e and the speed of its chang, $\dot{e}$, are also of the same value for any sample thickness. Thus according to Taylor's hypothesis there is a unique rheological relationship $R(e, \dot{e}, \sigma') = 0$ for all the samples of different thickness irrespective of the important differences in the effective stress speeds. These are very small too, but we have to bear in mind that even the primary consolidation of natural layers of the thickness of the order of magnitude 1 m or more occurs at such small effective stress speeds.

For a sufficiently small stress interval, $\Delta\sigma'$, the time lines corresponding to Taylor's Theory B represent, in the coordinate system (σ', e), a family of parallel straight lines whose equation is

$$e = C - a\sigma' + c\exp(-t/b) \tag{2.2}$$

Differentiation of this equation yields

$$e = C - a\sigma' - b\dot{e} - ab\dot{\sigma}' \tag{2.3}$$

With the substitution

$$C = e_o, \quad -a = B, \quad -b = A, \quad -ab = C \tag{2.4}$$

Eq. (2.3) represents rheological relationships for the rheological body according to Fig. 2.1 in case that $C = -a\,b = -A\,B$, i.e. for a special case of the Theory A of Taylor and Merchant (1940). If, however, according to the above mentioned hypothesis of the Theory B, one denies the influence of the speed $\dot{\sigma}'$ on the rheological relationships, Eq. (2.3) reduces to the basic equation of isotaches of the Kelvin body corresponding to the assumption of the Theory B:

$$e = C - a\sigma' - b\dot{e} \tag{2.5}$$

When in the coordinate system ($\log_{10} \sigma'/\sigma_0$, $\log_{10} e/e_0$) the time lines are equidistant straight lines (Hansen, 1969, Fig. 2.3), they can be expressed by the equation:

$$\frac{e}{e_0} = \left(\frac{\sigma'}{\sigma_0}\right)^{-\frac{c}{d}} \left(\frac{t_i + t}{t_0}\right)^{-c} \tag{2.6}$$

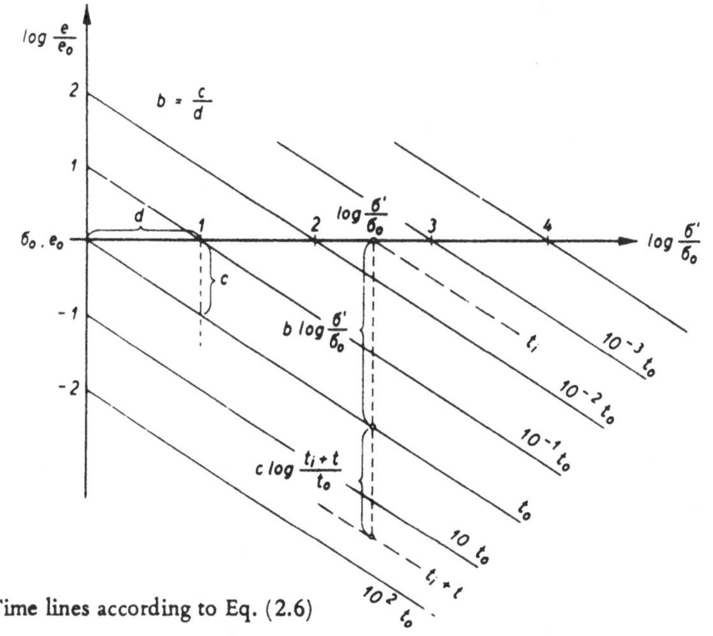

Fig. 2.3 - Time lines according to Eq. (2.6)

The differentiation with respect to the time t yields (Šuklje, 1975):

$$-\dot{e} = c\,e\,\left\{A\,e^{\frac{1}{c}}\,\sigma'^{\frac{1}{d}} + \frac{1}{d}\,\frac{1}{\sigma'}\,\dot{\sigma}'\left[1 - \left(\frac{e}{e_0}\right)^{\frac{1}{c}}\right]\right\} \tag{2.7}$$

where

$$A = \left(e_0^{\frac{1}{c}}\,\sigma_0^{\frac{1}{d}}\,t_0\right)^{-1} \tag{2.8}$$

Equation (2.7) represents an example of possible rheological relationships of the type $R(e, \dot{e}, \sigma', \dot{\sigma}') = 0$ (see also Garlanger, 1972).

If, according to Taylor's hypothesis, one neglects in Eq. (2.7) the second term including the factor $\dot{\sigma}'$, the reduced form of this equation

$$-\frac{\dot{e}}{\dot{e}_0} = \left(\frac{e}{e_0}\right)^{\frac{1+c}{c}} \left(\frac{\sigma'}{\sigma_0}\right)^{\frac{1}{d}} \tag{2.9}$$

represents a set of isotaches being parallel straight lines in the coordinate system ($\ln \sigma'/\sigma_0$, $\ln e/e_0$):

$$\ln\frac{e}{e_0} = -\left(\frac{1}{d}\,\frac{c}{1+c}\right)\ln\frac{\sigma'}{\sigma_0} - \frac{c}{1+c}\ln\frac{\dot{e}_0}{\dot{e}} \tag{2.10-a}$$

whereby

$$-\dot{e}_0 = \frac{c\,e_0}{t_0} \tag{2.10-b}$$

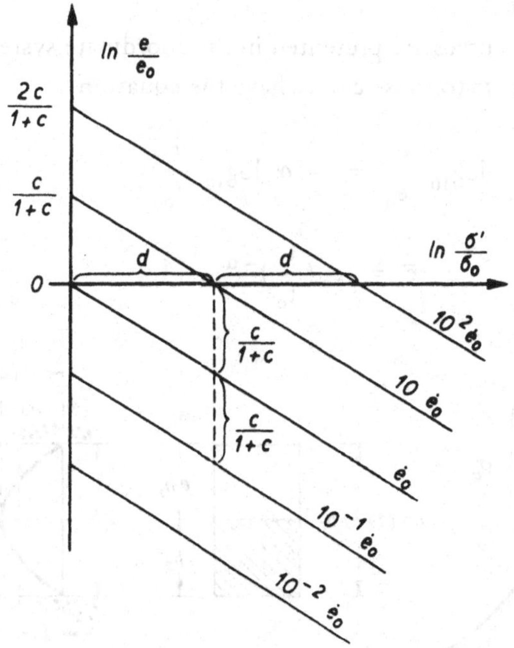

Fig. 2.4 - Isotaches according to Eq. (2.9)

Experimental Investigation of the Influence of the Speed $\dot{\sigma}'$ on the Rheological Relationships

From the consolidation curves of the oedometer test of a saturated sample, groups of data on e, $\dot{e}$, σ' et $\dot{\sigma}'$ may be deduced in the following approximate way.

The consolidation curves $\bar{e} = \bar{e}(t)$ (Fig. 2.5) represent the change of the average value of void ratios $\bar{e}$ with the time t. The tangents to these curves determine the mean values of the speed of the void ratio change $\dot{\bar{e}} = \sigma\bar{e}/\partial t$ which, in the secondary consolidation phase, represent the speeds $\dot{e} = \partial e/\partial t$ with a sufficient accuracy. If the consolidation curves are presented in the coordinate system $(\log_{10}(t/t_o), \bar{e})$, the consolidation speed equals

$$\dot{\bar{e}} = \frac{\partial \bar{e}}{\partial t} = -\frac{\vartheta_e}{Nt} \tag{2.11}$$

$$N = \frac{\ln x}{\log_{10} x} = 2.3026\,, \qquad \text{for } \vartheta_e \text{ see Fig. 2.4.}$$

When the consolidation curves are presented in the coordinate system $/\log_{10}(t/t_o)$, $\log_{10}(e/e_o)/$, the tangents to these curves have the equation

$$\log_{10}\frac{\bar{e}}{e_0} = -\alpha \log_{10}\frac{t}{t_o}$$

or

$$\bar{e} = e_o\left(\frac{t}{t_o}\right)^{-\alpha}$$

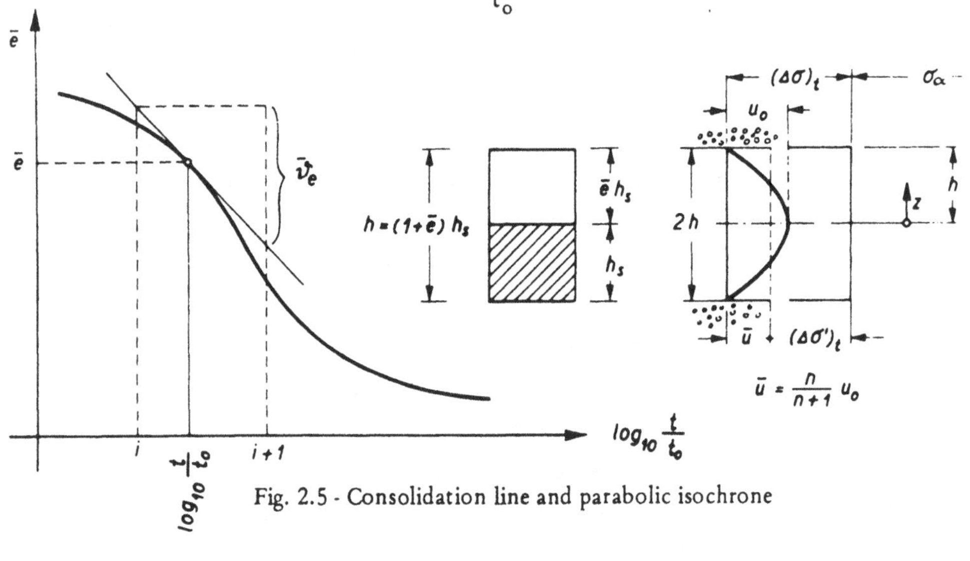

Fig. 2.5 - Consolidation line and parabolic isochrone

and their differentiation gives

$$\dot{\bar{e}} = -\frac{\alpha \bar{e}}{t} \tag{2.12}$$

By assuming a certain form of isochrones and the validity of Darcy's law $v = k\,i$, the mean value of the excess pore-pressure, $\bar{u}$, can be ascertained by equalizing the speed of the volume change and the seepage speed at the drained boundary of the sample (see Fig. 2.4). For the parabolic form of isochrones of the degree n

$$u = u_o\,[1 - (\frac{z}{h})^n] \tag{2.13}$$

we obtain the equation

$$-\,h_s \frac{\partial \bar{e}}{\partial t} = (k\,i)_{z=h} = \frac{1}{\gamma_w}\,(-k\,\frac{\partial u}{\partial z})_{z=h} = \frac{k_{z=h}}{\gamma_w}\,\frac{n u_o}{h}$$

(for notation see Fig. 2.5). By considering the relation

$$\bar{u} = \frac{n}{1+n}\,u_o \tag{2.14}$$

the above equation may be written:

$$\bar{u} = \frac{\gamma_w h_s^2}{1+n} \cdot \frac{1+\bar{e}}{k_{z=h}}\,\dot{\bar{e}} \tag{2.15}$$

h_s being the thickness of the solid constituent of the soil if assumed compact:

$$h_s = \frac{h}{1+\bar{e}}$$

The complementary value of the effective pressure $\bar{\sigma}'$ is defined by the equation

$$\bar{\sigma}' = \sigma - \bar{u}$$

and the speed of its change during the secondary consolidation of the oedometer sample when the excess pore-pressures become immeasurably small, can be taken equal to the speed of the effective pressure itself:

$$\frac{\partial \bar{\sigma}'}{\partial t} = - \frac{\partial \bar{u}}{\partial t} \cong \frac{\partial \sigma'}{\partial t}$$

By substituting $\bar{e} \cong e$, $k_{z=h} \cong k$, and taking into account that $e = e(t)$, $k = k(e)$, the differentiation of Eq. (2.15) yields:

(a) when the secondary consolidation line is straight in the system (log t/t_o, e):

$$(2.16) \qquad \frac{\partial \sigma'}{\partial t} = \frac{h_s^2 \gamma_w}{N^2(n+1)} \cdot \vartheta_e \cdot \frac{1}{(1+e)^2} \cdot \{ \vartheta_e + (1+e)[N - \frac{\vartheta_e}{k} \cdot \frac{\partial k}{\partial e}]\}$$

(b) when the secondary consolidation line is straight in the system (log t/t_o, log e/e_o):

$$(2.17) \qquad \frac{\partial \sigma'}{\partial t} = \frac{h_s^2 \gamma_w}{\ldots 1} \cdot \frac{\alpha e}{\ldots t} \cdot \{\alpha(1+2e) + (1+e)[1 - \frac{\alpha e}{k} \cdot \frac{\partial k}{\partial e}]\}$$

The function $k = k(e)$ can be established by the oedometer test itself. Usually, one of the following relations results:

$$(2.18) \qquad \begin{cases} \text{(i)} & k = k_o + Ae \\ \text{(ii)} & k = k_o \exp(C + Be) \\ \text{(iii)} & k = \sum_{i=0}^{n} (A_i e^i), \quad n > 1 \end{cases}$$

For partly saturated soils, the permeability depends also on the saturation degree S_r :

$$k = k(e, S_r) \tag{2.10}$$

Thus, in the domain of speeds $\dot{e}$ and $\dot{\sigma}'$ occurring during the secondary consolidation of oedometer samples, groups of simultaneous values e, $\dot{e}$, σ', $\dot{\sigma}'$ can be obtained. However, since for samples of equal thickness h_s any couple of e and σ' is associated with a single value of $\dot{e}$ and of $\dot{\sigma}'$, the equation $R(e, \dot{e}, \sigma') = 0$ fully defines the stress-strain-time relationship which can be deduced from conventional oedometer tests. On the other hand, the available experimental data from tests performed with samples of different thickness do not allow any clear conclusion on the effect of the speed $\dot{\sigma}'$ on the rheological relationships (Šuklje and Kovačič, 1974).

At the very small values of the speeds $\dot{\sigma}'$ appearing during the consolidation of layers whose thickness is of the order of magnitude 1m or more, the members containing $\dot{\sigma}'$ do not seem to influence the viscous behaviour of soils in an important manner. A comparative study of the consolidation of saturated layers performed alternatively by considering or neglecting these terms, has been proved favourable for such a hypothesis (Šuklje and Kovačič, 1974; see Chapter 10).

B. Viscous Rheological Models in Rotationally Symmetric State of Stress

3. A HYPERBOLIC STRESS-STRAIN RELATIONSHIP FOR FRICTIONLESS COHESIVE VISCOUS SOILS AT CONSTANT VOLUME

Singh and Mitchell (1968) have shown that the total strain $\dot{\epsilon}_T$ of many soils creeping under the influence of a constant deviator stress $(\sigma_1 - \sigma_3)$ may be expressed as a function of time t as

$$\dot{\epsilon}_T = x \exp(\alpha m) (\frac{t_1}{t})^{1-n} \tag{3.1}$$

where x, α , and n are constants to be determined, t_1 is unit time and m is the percentage of strength mobilized. For frictionless cohesive soils m is defined by the expression:

$$m = \frac{\sigma_1 - \sigma_3}{2c} \tag{3.2}$$

For values of $(1 - n)$ not equal to unity, the strain ϵ_T is determined by integration of Eq. (13.2):

$$\epsilon_T = x \exp(\alpha m) (t_1)^{1-n} n^{-1} t^n + \text{constant} \tag{3.3}$$

As at $t = 0$ the creep component of strain equals zero, the constant in Eq. (13.3) represents the instantaneous strain ϵ_i .

Palmerton (1972) assumed that ϵ_i obeys Kondner's equation (3.1) which may be written in the form

$$\sigma_1 - \sigma_3 = \frac{\epsilon_i}{\frac{1}{E_i} + \frac{\epsilon_i}{(\sigma_1 - \sigma_3)_{ult}}} \tag{3.4}$$

Furthermore, he assumed that the soil fails at a definite strain value ϵ_o . Thus, for frictionless cohesive soils we have, at $\epsilon_i = \epsilon_o$:

$$\sigma_1 - \sigma_3 = 2c$$

and Eq. (5.4) yields:

$$(\sigma_1 - \sigma_3)_{ult} = \frac{2cE_i\epsilon_o}{\epsilon_o E_i - 2c}$$

In this way, Eq. (3.4) can be written in the form

$$\epsilon_i = \frac{(\sigma_1 - \sigma_3)\bar{a}}{2c - (1 - \frac{\bar{a}}{\epsilon_o})(\sigma_1 - \sigma_3)} \qquad (3.5)$$

where

$$\bar{a} = \frac{2c}{E_i}$$

E_i being the initial tangent modulus. By inserting Eq. (3.5) for the constant in Eq. (3.3) and taking $t_1 = 1$, and

$$\frac{x}{n} = A$$

we get for the total strain ϵ_T the expression:

$$\epsilon_T = \frac{\bar{a}m}{1 - (1 - \frac{\bar{a}}{\epsilon_o})m} + A \exp(\alpha m)\, t^n \qquad (3.6)$$

The tangent modulus E_t is determined by calculating

$$E_t = \frac{d(\sigma_1 - \sigma_3)}{d\epsilon_T}$$

from Eq. (3.6) and is given by

$$E_t = \frac{2c[1-(1-\frac{\bar{a}}{\epsilon_o})m]^2}{\bar{a} + \{\alpha A \exp(\alpha m)\, t^n [1+(\frac{\bar{a}}{\epsilon_o}-1)m]^2\}} \qquad (3.7)$$

"The value of m within each element used for calculating E_t in the preceding equation is the average of the mobilized strength before and after the load is applied. (The first iteration uses the values of m as determined before the load application. Then after the new stress state is determined, the average values of m before and after load application is used in Eq. 3.7.) The creep movements are then computed at the desired times assuming that the mobilization factor m remains constant during creep. Thus an additional finite element computation is necessary to compute the movements at each time the result is desired and the value of the modulus used at each time step is calculated by Eq. (3.7)" (Palmerton, 1972).

"If the loads are to be applied to the structure at various times (as if following a construction sequence) then the creep movements from each separate load application may be added if it is assumed that the principle of superposition is valid". (Ibidem.)

4. STRESS-STRAIN-TIME RELATIONSHIPS IN TERMS OF OCTAHEDRAL VALUES

Soil as a Non-Linear Kelvin Body

If the stress and strain tensors are coaxial, the invariant octahedral values of stresses and strains can be expressed by principal stresses and strains resp. as follows:

$$\sigma_{oct} \equiv \sigma^o = \frac{1}{3}(\sigma_1 + \sigma_2 + \sigma_3) \tag{4.1}$$

$$\tau_{oct} \equiv \tau^o = \frac{1}{3}\sqrt{[(\sigma_1 - \sigma_2)^2 + (\sigma_2 - \sigma_3)^2 + (\sigma_3 - \sigma_1)^2]} \tag{4.2}$$

$$\epsilon_{oct} \equiv \epsilon^o = \frac{1}{3}(\epsilon_1 + \epsilon_2 + \epsilon_3) \tag{4.3}$$

$$\gamma_{oct} \equiv \gamma^o = \frac{2}{3}\sqrt{[(\epsilon_1 - \epsilon_2)^2 + (\epsilon_2 - \epsilon_3)^2 + (\epsilon_3 - \epsilon_1)^2]} \tag{4.4}$$

In triaxial tests of cylindrical samples $\sigma_2 = \sigma_3$ and $\epsilon_1 = \epsilon_3$. Assuming that the strains are small, we can take the volume strain ϵ_v to be equal to the sum of linear strains: $\epsilon_v = \epsilon_1 + 2\epsilon_3$. Consequently $\epsilon_3 = 1/2(\epsilon_v - \epsilon_1)$. The axial strain ϵ_1 and the volume strain ϵ_v can be measured. The octahedral values of stresses and strains are then given by the relations:

$$\sigma^o = \frac{1}{3}(\sigma_1 + 2\sigma_3) \qquad \tau^o = \frac{2}{3}(\sigma_1 - \sigma_3) \qquad (4.5), (4.6)$$

$$\epsilon^o = \frac{\epsilon_v}{3} \qquad \gamma^o = \frac{2}{3}(3\epsilon_1 - \epsilon_v) \qquad (4.7), (4.8)$$

In the secondary phase of the consolidation of triaxial cylindrical samples the creep can be expressed, in a wide domain of observations and possible extrapolations, by the equations:

$$\epsilon^o = \epsilon^o_0 + \epsilon_K[1 - (t/t_0)^m] \tag{4.9}$$

$$\gamma^o = \gamma^o_0 + \gamma_K[1 - (t/t_0)^n] \tag{4.10}$$

ϵ_0^o and γ_0^o denote octahedral values of linear and shear strains resp. at time t_0. All parameters appearing in Eqs. (4.9) and (4.10) are stress-dependent. In the conditions of drained triaxial tests the stress state is completely defined by octahedral effective stresses $\sigma^{o\prime}$ and τ^o. Thus

$$(4.11), (4.14) \qquad \epsilon_0^o = \epsilon_0^o(\sigma^{o\prime}, \tau^o) \qquad \gamma_0^o = \gamma_0^o(\tau^o, \sigma^{o\prime})$$

$$(4.12), (4.15) \qquad \epsilon_K = \epsilon_K(\sigma^{o\prime}, \tau^o) \qquad \gamma_K = \gamma_K(\tau^o, \sigma^{o\prime})$$

$$(4.13), (4.16) \qquad m = m(\sigma^{o\prime}, \tau^o) \qquad n = n(\tau^o, \sigma^{o\prime})$$

In special cases the above relationships can be given analytical expressions. In order to meet any possible variety of experimental data, they can be presented by sets of point values; the interpolation can be made by using splines (Desai, 1971).

The relations (4.9) and (4.10) may be differentiated with respect to $\sigma^{o\prime}$ and τ^o resp.:

$$(4.17) \qquad \frac{d\epsilon^o}{d\sigma^{o\prime}} = \frac{d\epsilon_0^o}{d\sigma^{o\prime}} + \frac{d\epsilon_K}{d\sigma^{o\prime}} \{1 - (t/t_0)^m\} - \epsilon_K (t/t_0)^m \ln(t/t_0) \frac{dm}{d\sigma^{o\prime}}$$

$$(4.18) \qquad \frac{d\gamma^o}{d\tau^o} = \frac{d\gamma_0^o}{d\tau^o} + \frac{d\gamma_K}{d\tau^o} \{1 - (t/t_0)^n\} - \gamma_K (t/t_0)^n \ln(t/t_0) \frac{dn}{d\tau^o}$$

By introducing the symbols:

$$(4.19), (4.20) \qquad \frac{d\sigma^{o\prime}}{d\epsilon^o} = 3K_t \qquad \frac{d\sigma^{o\prime}}{d\epsilon_0^o} = 3K_0$$

$$(4.21), (4.22) \qquad \frac{d\sigma^{o\prime}}{d\epsilon_K} = -3E_K \qquad \frac{dm}{d\sigma^{o\prime}} = B_o$$

$$(4.23), (4.24) \qquad \frac{d\tau^o}{d\gamma^o} = G_t \qquad \frac{d\tau^o}{d\gamma_0^o} = G_0$$

$$\frac{d\tau^o}{d\gamma_K} = -\,G_K \qquad\qquad \frac{dn}{d\tau^o} = B_\gamma \qquad (4.25),\ (4.26)$$

the tangential values K_t and G_t of the compression and shear modulus resp. can be computed by relations:

$$K_t = \frac{K_0}{1 - K_0/E_K + (t/t_0)^m\,\{K_0/E_K - 3\,K_0\,B_0\,\epsilon_K\,\ln(t/t_0)\}} \qquad (4.27)$$

$$G_t = \frac{G_0}{1 - G_0/G_K + (t/t_0)^n\,\{G_0/G_K - G_0\,B_\gamma\,\gamma_K\,\ln(t/t_0)\}} \qquad (4.28)$$

The differentiation of Eqs. (4.9) and (4.10) with respect to time yields:

$$\dot\epsilon^o = (-\,m/t_0)\,\epsilon_K\,(t/t_0)^{m-1} \qquad (4.29)$$

$$\dot\gamma^o = (-\,n/t_0)\,\gamma_K\,(t/t_0)^{n-1} \qquad (4.30)$$

At $t = t_0$ the strain speeds have the values $\dot\epsilon^o_0$ and $\dot\gamma^o_0$

$$\dot\epsilon^o_0 = (-\,m/t_0)\,\epsilon_K \qquad\qquad \dot\gamma^o_0 = (-\,n/t_0)\,\gamma_K \qquad (4.31),\ (4.32)$$

By dividing Eqs. (4.29) and (4.30) by Eqs. (4.31) and (4.32) resp., we get:

$$\dot\epsilon^o/\dot\epsilon^o_0 = (t/t_0)^{m-1} \qquad\qquad t/t_0 = (\dot\epsilon^o/\dot\epsilon^o_0)^{1/(m-1)} \qquad (4.33)$$

(4.34) $\dot{\gamma}^o/\dot{\gamma}^o_0 = (t/t_0)^{n-1}$ $\qquad t/t_0 = (\dot{\gamma}^o/\dot{\gamma}^o_0)^{1/(n-1)}$

If the speeds $\dot{\epsilon}^o_0$ and $\dot{\gamma}^o_0$ are related to the same reference time t_0, Eqs. (4.33) and (4.34) yield:

(4.35) $\dot{\epsilon}^o/\dot{\epsilon}^o_0 = (\dot{\gamma}^o/\dot{\gamma}^o_0)^{(m-1)/(n-1)}$

By inserting expressions (4.33) and (4.34) into Eqs. (4.9) and (4.10) we obtain the relations:

(4.36) $\epsilon^o = \epsilon^o_0 + \epsilon_K \{1 - (\dot{\epsilon}^o/\dot{\epsilon}^o_0)^\alpha\}$

(4.37) $\gamma^o = \gamma^o_0 + \gamma_K \{1 - (\dot{\gamma}^o/\dot{\gamma}^o_0)^\beta\}$

whereby the following substitutions have been made:

(4.38), (4.39) $\alpha = m/(m-1)$ $\qquad \beta = n/(n-1)$

Taking into account relations (4.11) to (4.16), Eqs. (4.36) and (4.37) represent a general form of rheological relationships corresponding to the non-linear Kelvin body.

Power Law of Undrained Tests

For undrained tests $\epsilon_v = 0$ and $\epsilon_2 = \epsilon_3 = -\epsilon_1/2$. Thus:

(4.40) $\gamma^o = \sqrt{2}\,\epsilon_1$ $\qquad \epsilon_1 = \gamma^o/\sqrt{2}$

and according to Eq. (4.30):

$$\dot{\epsilon}_1 = (-n/\sqrt{2})\ (\gamma_K/t_0)\ (t/t_0)^{n-1} \tag{4.41}$$

By substituting

$$(-n/\sqrt{2})\ (\gamma_K/t_0) = \Phi(\tau^o, \sigma^{o\prime}) = x\ \exp(\alpha\, m)$$

$$m = (\sigma_1 - \sigma_3)/2c = (3/4)\sqrt{2}\ \tau^o/c\ , \qquad t_0 \equiv t_1$$

Eq. (4.11) gets the form (3.1) of the power law of Singh and Mitchell (1968) accounting for cohesive soils ($\phi = 0$):

$$\dot{\epsilon}_1 = x\ \exp(\alpha\, m)\ (t_1/t)^{1-n} \tag{3.1}$$

Isotaches

For certain constant strain-speed ratios $\dot{\epsilon}^o/\dot{\epsilon}^o_0$ and $\dot{\gamma}^o/\dot{\gamma}^o_0$ resp. the expressions (4.36) and (4.37) represent families of isotaches, i.e. strain versus stress plots corresponding to selected constant strain speeds:

$$\epsilon^o = \{\epsilon^o(\sigma^{o\prime}, \tau^o)\}_{\dot{\epsilon}^o = \mathrm{const}_\epsilon} = \{[\epsilon^o(\sigma^{o\prime})]_{\tau^o = \mathrm{const}_\tau}\}_{\dot{\epsilon}^o = \mathrm{const}_\epsilon} \tag{4.42}$$

$$\gamma^o = \{\gamma^o(\tau^o, \sigma^{o\prime},)\}_{\dot{\gamma}^o = \mathrm{const}_\gamma} = \{[\gamma^o(\tau^o)]_{\sigma^{o\prime} = \mathrm{const}_\sigma}\}_{\dot{\gamma}^o = \mathrm{const}_\gamma} \tag{4.43}$$

For a given stress state the distance between two adjacent isotaches (4.42) corresponding to the strain-speed ratios

$$\dot{\epsilon}^o/\dot{\epsilon}^o_0 = 10^{-(i-1)} \qquad \text{and} \qquad 10^{-i}, \qquad i = 1,2,3,\ldots$$

can be obtained by subtracting the corresponding ϵ^o values according to Eq. (4.36):

$$(4.44) \qquad \Delta\epsilon^o_i = \epsilon^o_i - \epsilon^o_{i-1} = \epsilon_K(1 - 10^{-\alpha})\ 10^{-(i-1)}$$

$$(4.45) \qquad \epsilon_K(1 - 10^{-\alpha}) = \Delta\epsilon_{i=1}$$

$$(4.46) \qquad \Delta\epsilon^o_i / \Delta\epsilon^o_{i=1} = 10^{-(i-1)\alpha}$$

In an analogous way, the distance between two adjacent isotaches (4.44) can be obtained by the relation

$$(4.47) \qquad \Delta\gamma^o_i / \Delta\gamma_{i=1} = 10^{-(i-1)\beta}$$

At $\alpha = 1$ the distance between the isotaches (4.43) decrease. The decrease is the more important the greater the parameter α. For $\alpha = 1$ the distance at $i = 3$, i.e. between the adjacent isotaches $i = 2$ and $i = 3$, reduces to one hundredth of the first distance $\Delta\epsilon_{i=1}$ and at smaller strain speeds the viscous effects practically disappear (Šuklje and Kovačič, 1978).

If ϵ^o_0 is a linear function of stresses, ϵ_K = const, and $\alpha = 1$, the rheological model corresponds to the linear Kelvin body. The obtained simplified relationship is not able to express real soil properties except in special cases and for a very limited stress interval.

At $\alpha \to 0$ the distance between subsequent isotaches become equal and Eq. (4.36) has to be substituted by the equation:

$$(4.48) \qquad \epsilon^o = \epsilon^o_0 + \beta^o \ \ln(\dot{\epsilon}^o_0/\dot{\epsilon}^o)$$

$$(4.48\text{-a, b}) \qquad \epsilon^o_0 = \epsilon^o_0(\sigma^{o\prime}, \tau^o), \qquad \beta^o = \beta^o(\sigma^{o\prime}, \tau^o)$$

For $\beta > 0$ analogous conclusions can be deduced for the intervals between the isotaches (4.44), and for $\beta \to 0$ Eq. (4.37) has to be replaced by the relationship

$$\gamma^o = \gamma_0^o + \beta_\gamma \ \ln(\dot{\gamma}_0^o/\dot{\gamma}^o) \tag{4.49}$$

$$\gamma_0^o = \gamma_0^o(\tau^o, \sigma^{o\prime}) \ , \qquad \beta_\gamma = \beta_\gamma(\tau^o, \sigma^{o\prime}) \qquad \text{(4.49-a,b)}$$

For $\beta < 0$ the distances between subsequent isotaches (4.44) increase indicating the approach of failure: Mohr's stress circles are close to the strength envelope.

In the coordinate system $(\epsilon^o, \log_{10}(t/t_0))$, the tangent to the secondary consolidation line corresponding to a selected stress state $(\sigma^{o\prime}, \tau^o)$ has, at $t = t_0$, the equation

$$\epsilon^o = \epsilon_0^o + \beta_o \ \ln(t/t_0) = \epsilon_0^o + \beta_o \ N \log_{10}(t/t_0) \tag{4.50}$$

(Fig. 4.1). If in the same coordinate system the secondary consolidation line according to Eq. (4.9) is plotted, the strain ϵ^o increases, along this line, in a logarithmic time unit $\Delta(t/t_0) = 10^i - 10^{i-1}$ by

$$\Delta\epsilon_{t,i}^o = \epsilon_K (1 - 10^m) \ 10^{(i-1)m} \tag{4.51}$$

Substituting

$$\epsilon_K (1 - 10^m) = \Delta\epsilon_{t,i=1}^o \tag{4.52}$$

we can write

$$\Delta\epsilon_{t,i}^o / \Delta\epsilon_{t,i=1}^o = 10^{(i-1)m} \qquad \text{(4.51-a)}$$

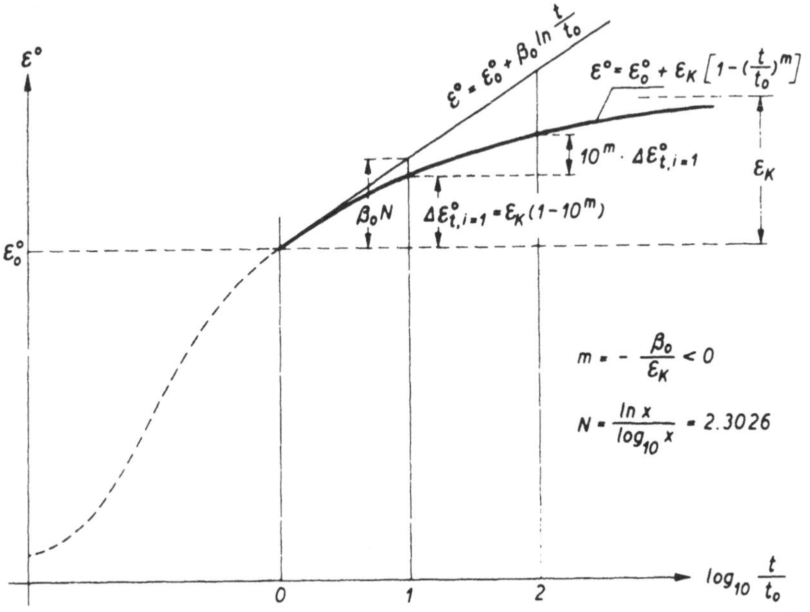

Fig. 4.1 - Deviation of the secondary consolidation line from the logarithmic straight line

Since in Eq. (4.9) the parameters m are negative, the increase of the value ϵ^o from $\epsilon^o = \epsilon^o_0$ at $t = t_0$ up to the asymptotic value at $t \to \infty$ equals ϵ_K (Fig. 4.1). Requiring, at $t = t_0$, equal values of differential quotients $d\epsilon^o/d(t/t_0)$ for the plots according to respective relations (4.9) and (4.50) we obtain the relation

$$-\epsilon_K\, m = \beta_o \tag{4.53}$$

At $m \to 0$ the parameter ϵ_K tends to infinity: the plot according to Eq. (4.9) approaches the logarithmic straight line according to Eq. (4.50).

Analogous conclusions can be made for the distortional constituents of the strain tensor.

Examples

(a) Drained Tests

In Figs. 4.2 to 4.5 an example of rheological relationships as ascertained by drained triaxial tests is presented. The soil is a marsh silty clay of high compressibility. During the laboratory observation the creep developed according to Buisman's law. The viscous effects have been extrapolated in the direction of the observed logarithmic straight lines of the secondary consolidation. Thus, the rheological relationships have been considered to have the form of Eqs. (4.48) and (4.49).

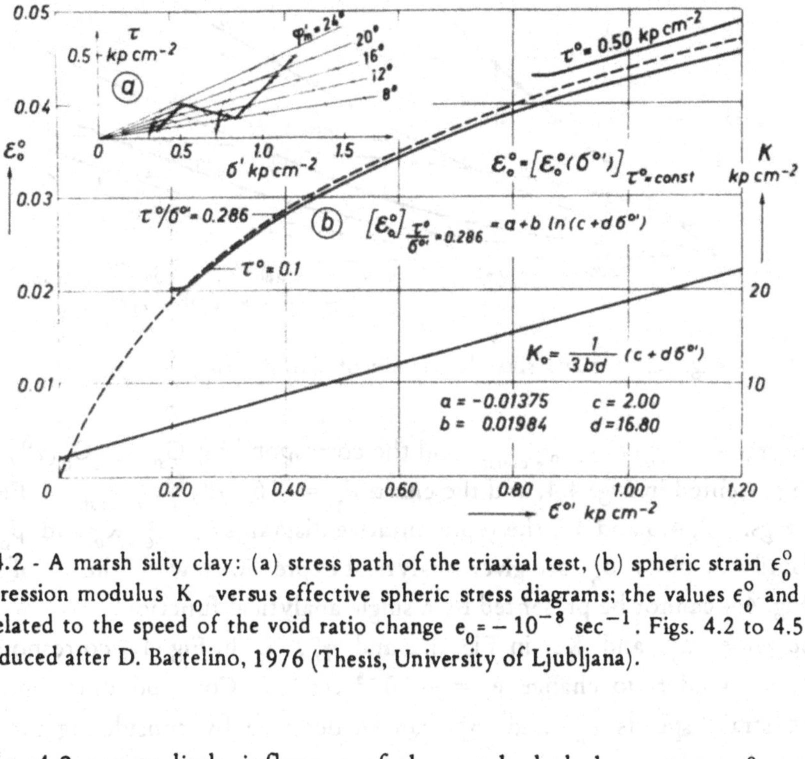

Fig. 4.2 - A marsh silty clay: (a) stress path of the triaxial test, (b) spheric strain ϵ_0^o and compression modulus K_0 versus effective spheric stress diagrams; the values ϵ_0^o and K_0 are related to the speed of the void ratio change $\dot{e}_0 = -10^{-8}\ \mathrm{sec}^{-1}$. Figs. 4.2 to 4.5 are reproduced after D. Battelino, 1976 (Thesis, University of Ljubljana).

Fig. 4.2 proves little influence of the octahedral shear stress τ^o onto the $\epsilon_0^o = \epsilon_0^o(\sigma^{o\prime})$ relationship. For this reason the plot $\epsilon_0^o = \{\epsilon_0^o(\sigma^{o\prime})\}_{\tau^o/\sigma^{o\prime} = 0.286}$ and the corresponding K_0 line (see Eq. 4.17) can be taken as representative for all stress states. Since in the case presented the spheric viscous effects (β_0, Fig. 4.3) were found to be small as compared to the ϵ_0^o values, the plot $\beta_0 = \{\beta_0(\sigma^{o\prime})\}_{\tau^o/\sigma^{o\prime} = 0.246}$ can be taken into account throughout the entire stress field.

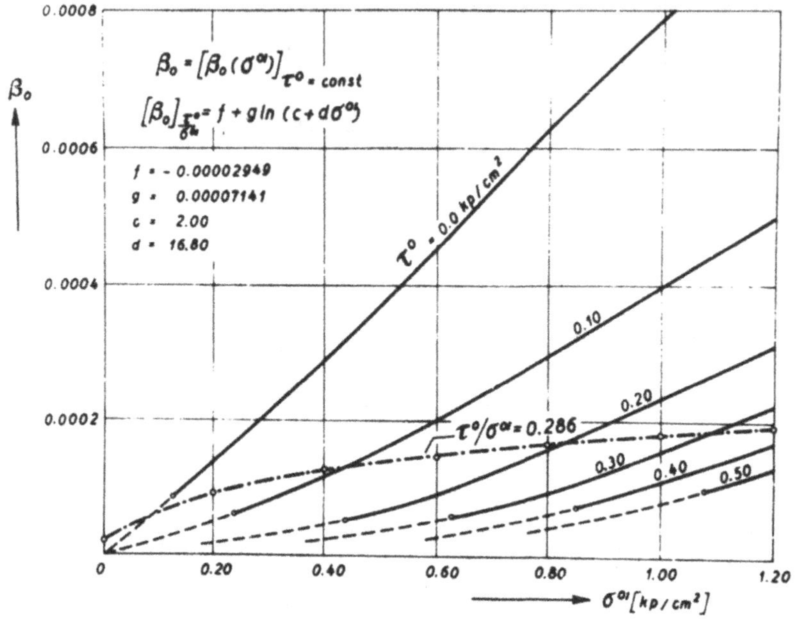

Fig. 4.3 - A marsh silty clay: family of plots $\beta_0 = \{\beta_0(\sigma^{0'})\}_{\tau^0 = \text{const.}}$

The $\gamma_0^o = \{\gamma_0^o(\tau^o)\}_{\sigma o' = \text{const}}$ and the corresponding $G_0 = \{G_0(\tau^o)\}_{\sigma o' = \text{const}}$ plots are presented in Fig. 4.4, and the charts $\beta_\gamma = \{\beta_\gamma(\tau^o)\}_{\sigma o' = \text{const}}$ in Fig. 4.5.

In Figs. 4.2, 4.3 and 4.5 the representative diagrams for ϵ_0^o, K_0 and β_0 as well as the family of plots β_γ are given analytical expressions, while the set of plots for γ_0^o in Fig. 4.4 cannot be presented by a single analytical function.

The values ϵ_0^o and K_0 in Fig. 4.2 and γ_0^o, G_0 in Fig. 4.4 correspond to the speed of the void ratio change $\dot{e}_0 = -10^{-8}$ sec^{-1}. Corresponding simultaneous reference strain speeds $\dot{\epsilon}_0^o$ and $\dot{\gamma}_0^o$ can be deduced by considering the relation (4.35) and the following connections:

(4.54-a,b) $$e = e_0 - 3\,\epsilon^o(1 + e_0) \rightarrow \dot{e} = -3\,\dot{\epsilon}^o(1 + e_0)$$

e_0 being the initial void ratio (corresponding to $\epsilon^o = 0$).

(b) Undrained Tests

Figs. 4.6 to 4.9 show the results of consolidated undrained tests of a marsh silt of high compressiblity. The $\gamma_0^o = \{\gamma_0^o(\tau^o)\}_{\sigma^o = \text{const}}$ plots in Fig. 4.6 and the corresponding $G_0 = \{G_0(\tau^o)\}_{\sigma^o = \text{const}}$ plots in Fig. 4.7 are related to the reference speed $\dot{\gamma}_0^o = -10^{-8}$ sec.$^{-1}$.

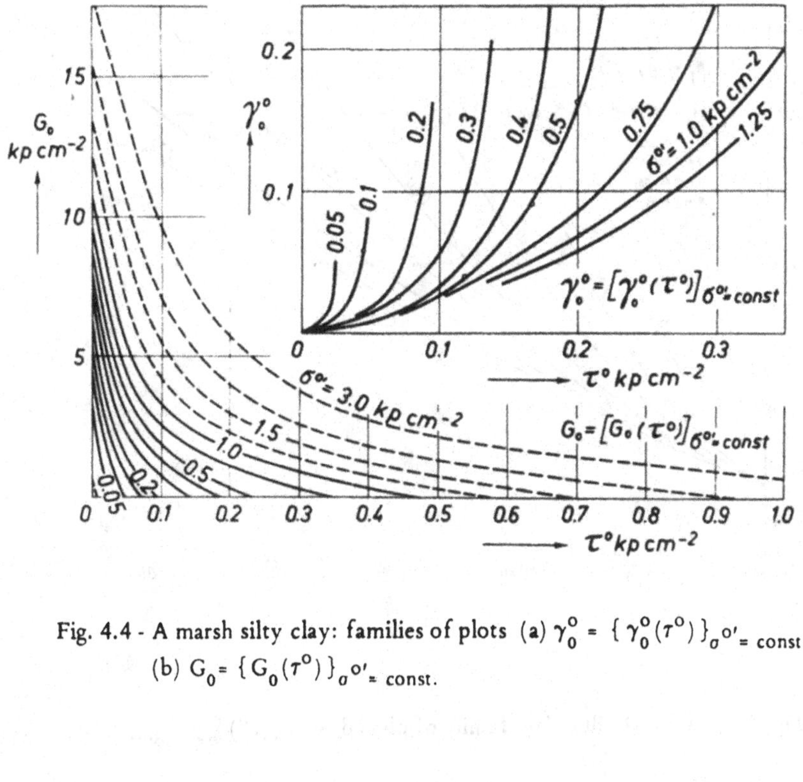

Fig. 4.4 - A marsh silty clay: families of plots (a) $\gamma_0^o = \{\gamma_0^o(\tau^o)\}_{\sigma^{o\prime} = \text{const}}$.
(b) $G_0 = \{G_0(\tau^o)\}_{\sigma^{o\prime} = \text{const}}$.

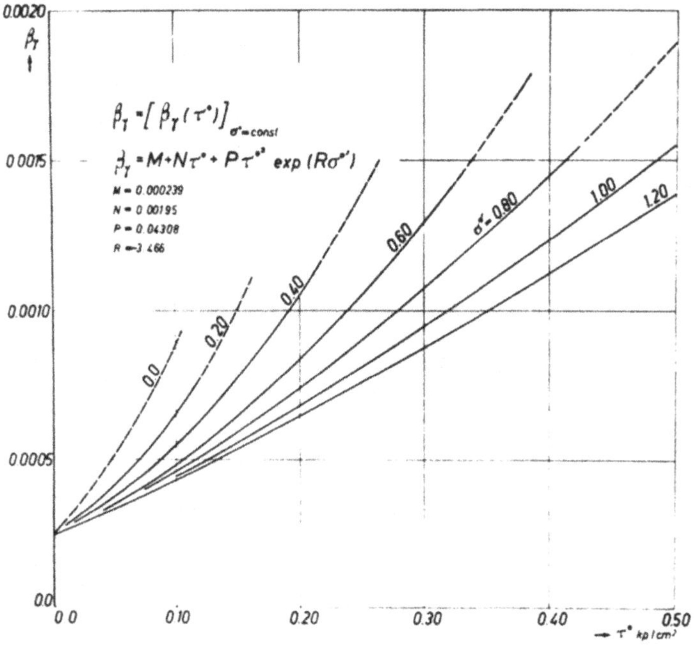

Fig. 4. 5 - A marsh silty clay: family of plots $\beta_\gamma = \{\beta_\gamma(\tau^0)\}_{\sigma^{\circ\prime} = \text{const.}}$

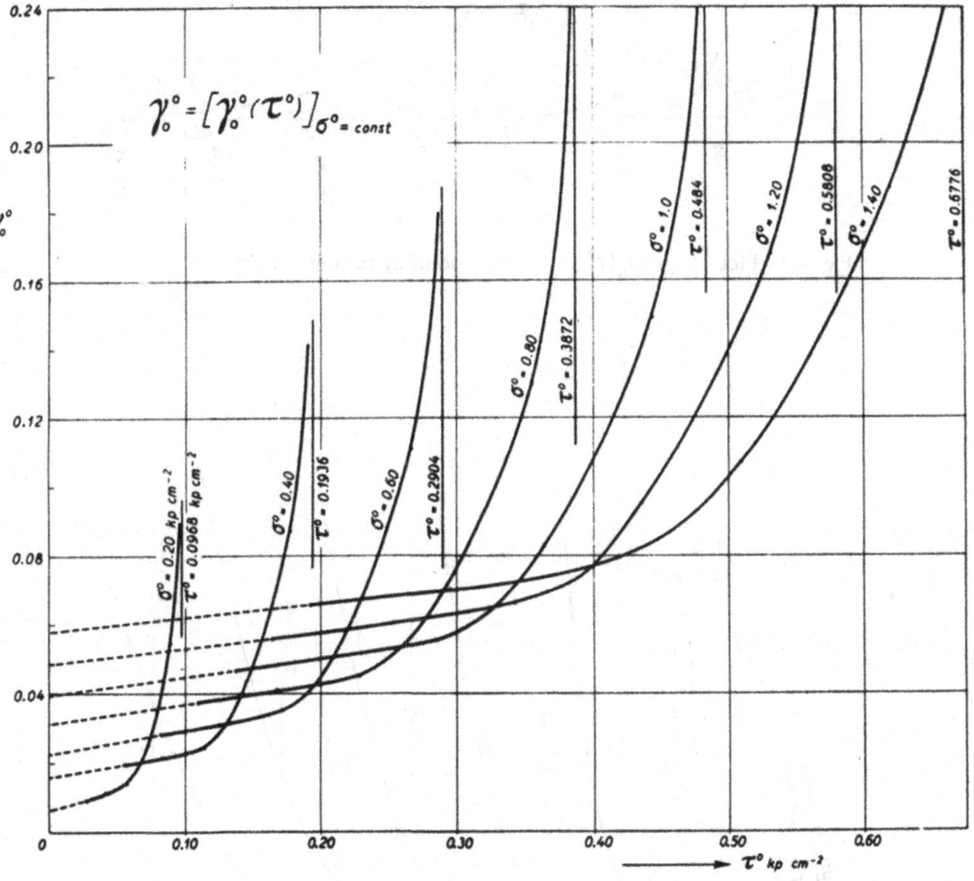

Fig. 4.6 - A marsh silt: Plots $\gamma_0^o = \gamma_0^o(\tau^o, \sigma^o)$

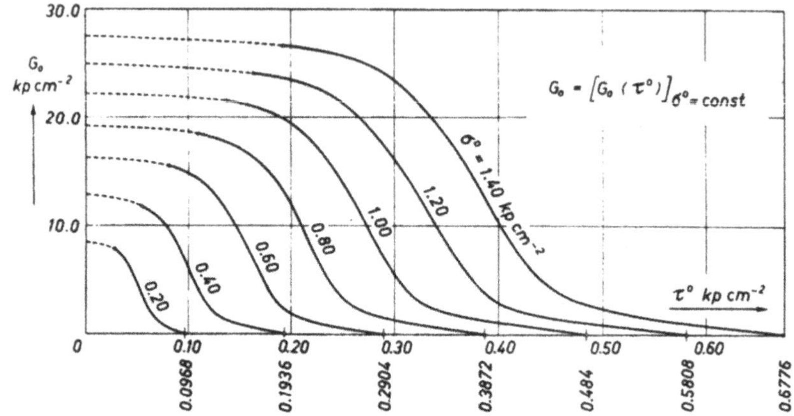

Fig. 4.7 - Plots $G_0 = G_0(\tau^o, \sigma^o)$ corresponding to plots in Fig. 4.6.

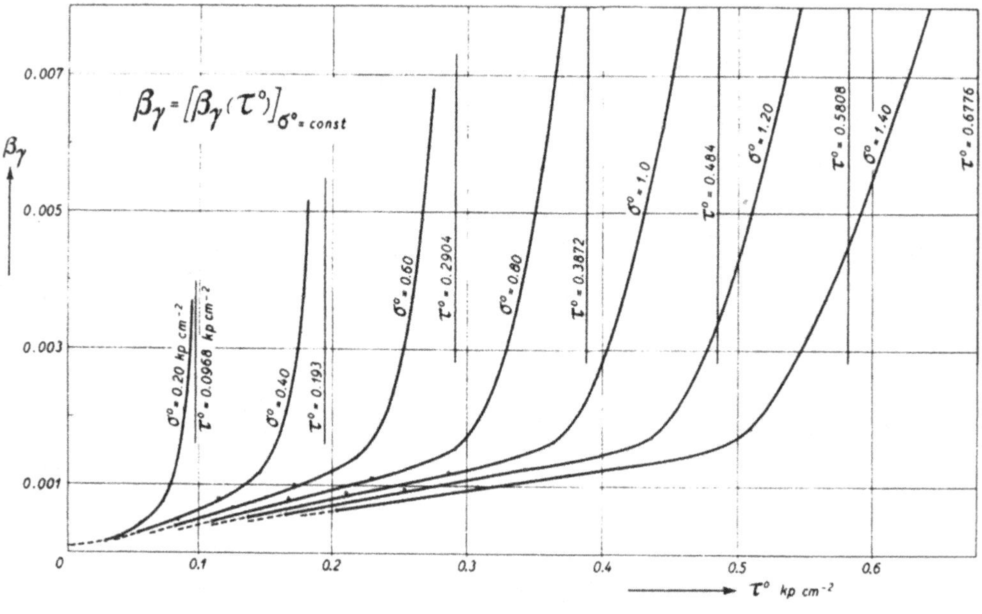

Fig. 4.8 - A marsh silt: Plots $\beta_\gamma = \beta_\gamma(\tau^o, \sigma^o)$.

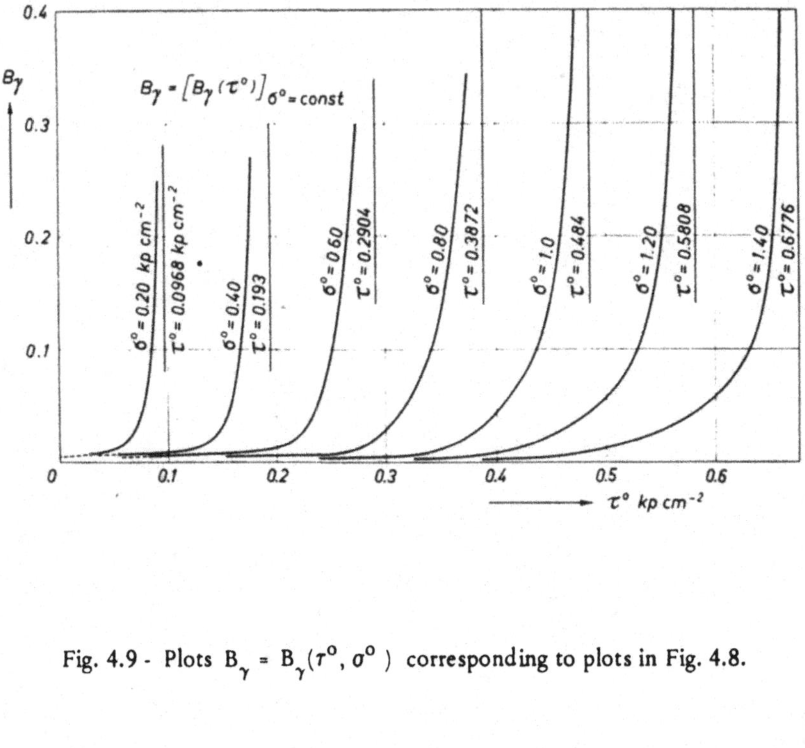

Fig. 4.9 - Plots $B_\gamma = B_\gamma(\tau^o, \sigma^o)$ corresponding to plots in Fig. 4.8.

C. Viscous Rheological Models for Arbitrary State of Stress

5. RELATIONSHIPS ACCORDING TO CERTAIN SIMPLE RHEOLOGICAL MODELS

The use of simple rheological models for expressing stress-strain-time relationships for soils can be justified if the properties of such models correspond to experimental observation. The existing verifications seem, however, to have been made for too limited stress, strain and time intervals. Their extrapolation and generalization can lead to eroneous conclusions.

The first application of a rheological model for expressing stress-strain-time-relationships in three-dimensional stress conditions was made by Tan (1954, 1957). His model consists of the Hookean spring connected in parallel with the Maxwell body: the spring is related to the spherical part and the Maxwell body to the deviatoric part of the stress-strain relationship (Fig. 5.1). Tan's rheological model can be considered as a special case of Anagnosti's (1962) generalized model in which Hooke's spring has been replaced by the Kelvin body.

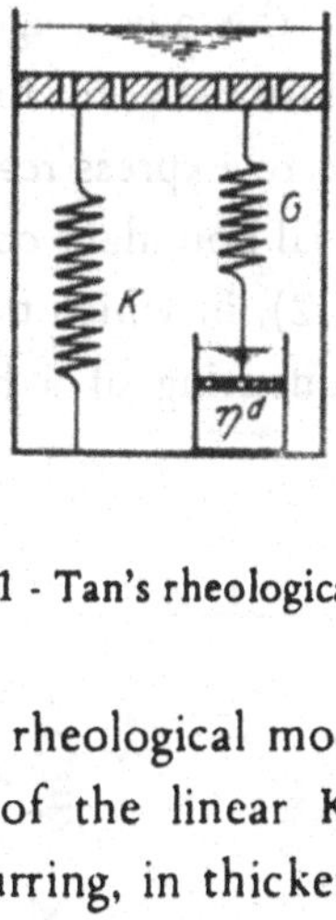

Fig. 5.1 - Tan's rheological model

When applying Anagnosti's rheological model to the conditions of linear strain states, the same shortcomings of the linear Kelvin model appear as in Taylor's Theory B: at small speeds occurring, in thicker layers, already during the primary consolidation, the viscous soil properties have hardly any effect onto the consolidation process. Even when the seepage paths are short, the secondary spheric strains tend rather early to their end values. As in linear strain conditions they are linearly proportional to the deviatoric strains, the deviator stress components dissipate. Consequently, the entire stress state tends to become hydrostatic. This conclusion does not agree with the observations and cannot be accepted.

For general space strain states, or for rotationally symmetric states treated by Anagnosti, the viscous effects, as represented by the selected rheological scheme, onto the spheric consolidation are analogous: When the seepage paths are long, the volume consolidation tends to the final state as early as at the end of the primary phase. The deviatoric strain components can, however, continue at constant effective stress states with constant speed corresponding to the law of the linear Maxwell body. In soils such strain progress can be observed only at the stress states approaching failure.

The solutions presented by Freudenthal and Spillers (1964) for a viscous elastic layer whose deformability corresponds to the linear Maxwell body, are subjected to analogous restrictions.

If, however, the linear Kelvin model was used for the spheric and deviatoric components of the rheological relationships, the entire consolidation of thicker layers tend to the final state already at the end of the primary phase, and viscous effects can hardly be observed. (Such a rheological model was applied by Zaretsky, 1967, in some concrete solutions.)

Linear rheological models have the advantage that the problems can rather simply be transformed into the solutions according to the theory of elasticity. Unfortunately, these models are able to express real soil behaviour only in limited stress and strain domains and for special boundary conditions. This limitation affects also the so-called M/V model (Fig. 5.2), in which the Maxwell body is connected in parallel with a two-element body consisting of a Hookean spring in series with a

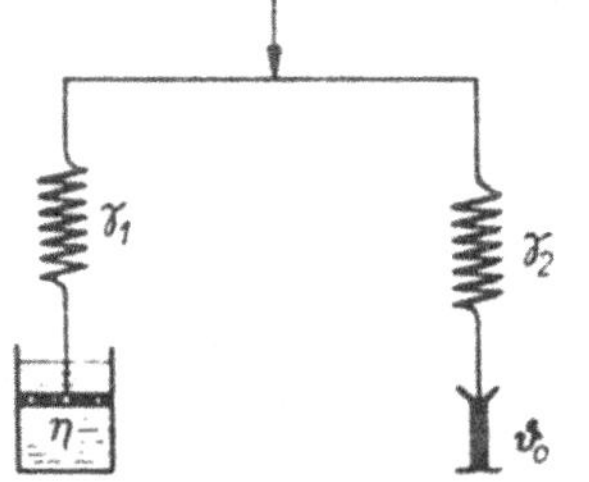

Fig. 5.2 - M/V rheological model

Saint-Venant resistance. Such a model was applied by Folque (1961) and was used by Kisiel and Lysik (1966 and in previous Kisiel's studies) when investigating the consolidation process in the half-space or half-plane subjected to surface loads; the corresponding rheological relationships had been combined with Biot's consolidation equation (Biot, 1935 and 1941).

The applicability of rheological models can be improved if they are composed of elements with non-linear parameters. In Stroganov's (1963) Kelvin model the non-linearity concerns the elastic moduli (E and G) while the coefficients of viscosity are kept constant. Consequently, when applying this model, we cannot expect any important role of viscous soil properties in the consolidation of layers with low permeability and long seepage paths. In this respect the application of models with non-linear viscous elements seems to be more promising. Murayama and Shibata (1961) used the connection in series of a Hookean spring E_1 with a modified Kelvin body containing in parallel connection a Hookean spring E_2, a non-linear Newton element η_2 and a Saint-Venant resistance σ_0 (Fig. 5.3). Nevertheless, the stress domain of the applicability of such a model remains very limited.

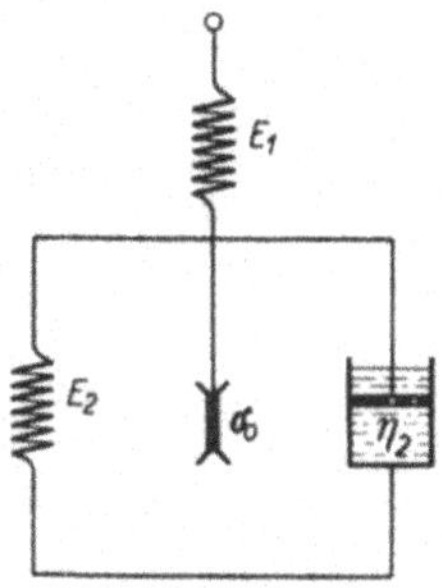

Fig. 5.3 - Murayama and Shibata's rheological model

The methods of numerical analyses facilitate in a large extent the use of complex rheological relationships. However, it does not seem reasonable to restrict the rheological properties as observed in laboratory and in nature, into the narrow frames of classical rheological models combined of single simple elements. Until general constitutive laws accounting for viscous soil properties are available, the experimental data are preferably to be expressed by graphical and analytical relationships or by single point coordinates in the form suitable for the use of computers in solving the problems by appropriate numerical procedures.

6. APPLICATION OF THE THEORY OF HEREDITARY CREEP

Vyalov (1963) suggested to present the results of experimental investigations of the deviatoric deformability in the coordinate system (σ^o, T, Γ) where

$$T = \frac{\sqrt{3}}{2}\tau^o , \qquad \Gamma = \frac{\sqrt{3}}{2}\gamma^o$$

T is the "intensity of shear stress" and Γ the "intensity of shear strain". Viscous effects have been expressed according to the theory of hereditary creep (Boltzman – Volterra). By taking into account the experimental data and by using some simplifying assumptions, the following stress-strain relationships have been suggested:

$$T = A(t)\Gamma^m + B(t)\sigma^o \Gamma^n \qquad (6.1)$$

The relaxation functions A(t) and B(t) can be approximated in the form:

$$A(t) = \frac{A}{1 + (at)^{\lambda_a}} , \qquad B(t) = \frac{B}{1 + (bt)^{\lambda_b}}$$

When $m = n$, $a = b$, $\lambda_a = \lambda_b = \lambda$, Eq. (6.1) yields:

$$T = \frac{A\Gamma^m}{1 + (at)^{\lambda}} \left(1 + \frac{\sigma^o}{H}\right) , \qquad H = \frac{A}{B} \qquad (6.2)$$

In a more general form the theory of the hereditary creep has been applied by Zaretsky (1967). Now, his rheological equations are not immediately related to effective stresses. He combines the deformations due to total stresses with the spheric dilatation due to the effect of the excess pore-pressures p(t).

According to Zaretsky, the stress tensors for viscous soils can be expressed in the following general form:

$$\sigma_{ij}(t) = 2\bar{G}\{\epsilon_{ij}(t)\} + \delta_{ij}[\bar{\lambda}\{\epsilon_v(t)\} - \frac{\bar{\alpha}_v}{\bar{\alpha}_{vp}}\{p(t)\}] \qquad (6.3)$$

(6.4) $$\sigma_{kk}(t) = I_1(t) = \bar{\alpha}_v\{\epsilon_v(t)\} - 3\frac{\bar{\alpha}_v}{\bar{\alpha}_{vp}}\{p(t)\}$$

with the following notation:

$$\bar{G}\{y(t)\} = G[y(t) - \int_0^t R_i(t,\tau)\, y(\tau)\, d\tau]$$

$$\bar{\lambda}\{y(t)\} = \frac{1}{3}[\bar{\alpha}_v\{y(t)\} - 2\bar{G}\{y(t)\}]$$

$$\bar{\alpha}_v\{y(t)\} = \alpha_v[y(t) - \int_0^t R_v(t,\tau)\, y(\tau)\, d\tau]$$

By adding the index p, equation for $\bar{\alpha}_v\ \{y(t)\}$ can be applied also for $\bar{\alpha}_{vp}$. α_v is the compression modulus, and α_{vp} the dilatation modulus of the mineral skeleton, $R_v(t,\tau)$ and $R_{vp}(t,\tau)$ are the resolvents of the kernel of the compression and dilatation creep respectively, and $R_i(t,\tau)$ the resolvent of the kernel of the distortional creep.

In Zaretsky's book (1967) concrete solutions of deformation and consolidation problems have been presented for the resolvents of the kernel which corresponds either to the linear Kelvin body, or of the kernel of creep having the form:

(6.5) $$K(t-\tau) = \frac{\delta}{(t-\tau)^{\gamma}}, \qquad 0 < \gamma < 1$$

including, with $\gamma = 1$, the logarithmic law of the secondary compression. Solutions have been given also for the combination of the relation (6.5) with the creep of a linear Kelvin body:

(6.7) $$K(t-\tau) = \frac{\delta \exp[-\delta_1(t-\tau)]}{(t-\tau)^{\gamma}}, \qquad 0 < \gamma < 1$$

Now, the compression due to total stresses, and the dilatation due to the effect of pore pressures do not occur successively, but simultaneously. Consequently, the separate treatment of compression and dilatation soil behaviour seems to be questionable. If, however, $\tilde{\alpha}_v = \tilde{\alpha}_{vp}$ and $R_v(t, \tau) = R_{vp}(t, \tau)$ the procedure of Zaretsky involves the principle of effective stress in the form as presented in the Chapter 1.

7. COMPLEX RHEOLOGICAL RELATIONSHIPS

Parameters appearing in rheological models dealt with in Chapters 3 and 4, have been deduced from the tests performed in conventional triaxial apparatuses, i.e. at rotationally symmetric state of stress. Now, similar stress-strain-time relationships can be obtained in plane-strain triaxial devices whose construction is rather simple. If results from tests made in such devices are available, the relationships expressed in terms of either principal or octahedral stresses can immediately be applied in calculating stresses, strains and their development with time in conditions of plane strain. If only results from testing cylindrical samples are available, appropriate simplifying assumptions have to be made, e.g. the assumption that the intermediate principal stress does not affect the rheological relationships.

A better approach to the real soil behaviour should be the subject of our future investigations. So far there are hardly any data from "real" triaxial apparatuses available concerning the viscous effects onto the deformability of soils in arbitrary states of stress. Rheological models presented in the two foregoing chapters, whose use has been suggested for arbitrary stress states, have not been checked for the influence of the intermediate principal stress either.

II. DEVELOPMENT OF STRESSES AND STRAINS IN VISCOUS SOILS

8. STRESSES AND STRAINS IN THE NON-LINEAR VISCOUS PLANE-STRAIN SPACE WITHOUT SEEPAGE RESISTANCE

Numerical Procedure

(a) Soil as a Non–Linear Kelvin Body

Let us assume that at the beginning of a time interval $\Delta t_i = t_i - t_{i-1}$ all displacements, stresses, strains and strain-speeds are known:

displacements $\{U\}_{i-1} = \begin{Bmatrix} u \\ v \end{Bmatrix}_{i-1}$

strains $\{\epsilon\}_{i-1} = \begin{Bmatrix} \epsilon_{xx} \\ \epsilon_{yy} \\ \epsilon_{xy} \end{Bmatrix}_{i-1}$ and the corresponding octahedral values ϵ^o_{i-1} and γ^o_{i-1}

stresses $\{\sigma'\}_{i-1} = \begin{Bmatrix} \sigma_{xx} \\ \sigma_{yy} \\ \tau_{xy} \end{Bmatrix}_{i-1}$ and the corresponding octahedral values $\sigma^{o\prime}_{i-1}$ and τ^o_{i-1}

octahedral strain speeds $\dot{\epsilon}^o_{i-1}$ and $\dot{\gamma}^o_{i-1}$.

Since according to Eqs. (4.33) and (4.34), the time ratio can be expressed by the speed ratio and since all parameters appearing in expressions (4.27) and (4.28) are known functions of octahedral stresses, the tangent moduli at time t_{i-1} can be obtained for all finite elements and the corresponding stiffness matrix $[K]_{i-1}$ can be determined. This matrix is then used for the computation of the first approximation of the displacement increments in nodal points of finite elements, corresponding to the increase of nodal forces by $\{\Delta P\}_i$:

$$\{\Delta U\}_i = [K]^{-1}_{i-1} \quad \{\Delta P\}_i \tag{8.1}$$

The corresponding strain and stress increments can be obtained from the relations

$$\{\Delta\epsilon\} = [p] \ \{\Delta U\} \tag{8.2}$$

and

$$(8.3) \qquad \{\Delta\sigma'\} = [C]\ \{\Delta\epsilon\}$$

The matrix [p] is the product of the matrix of differential operators, of the field matrix of the element, and of the inverse field matrix of boundary values, while [C] denotes the elastic matrix corresponding to the tangent moduli K_t and G_t at time t_{i-1}. By adding the obtained strain and stress increments to the previous strains and stresses, we obtain the first approximation for $\{\epsilon\}_i$ and $\{\sigma\}_i$ and for corresponding octahedral values governing the rheological parameters according to Eqs. (4.11) to (4.16) and (4.19) to (4.26).

Now, the real values of octahedral strains depend on the creep effect in the time interval Δt_i. If the reference time t_{0_i} and the relative time t_i at the end of the time interval Δt_i are known, the strains ϵ_i^o and γ_i^0 can be obtained from Eqs. (4.9) and (4.10). By using, for parameters m_i and ϵ_{K_i}, the first approximation, the reference time t_{0_i} can be computed from Eqs. (4.31) or (4.32):

$$(8.4) \qquad t_{0i} = -\,m_i\,(\epsilon_{Ki}/\dot{\epsilon}_0^o) \qquad -\,n_i\,(\gamma_{Ki}/\dot{\gamma}_0^o)$$

Equation (4.9) or (4.10) can now be applied to the strain ϵ_{i-1}^o or γ_{i-1}^o at the beginning of the time interval Δt_i, under the assumption that at t_{i-1} loads and stresses suddenly increase by $\{\Delta P\}_i$ and $\{\Delta\sigma\}_i$ respectively. The application of Eq. (4.9)

$$\epsilon_{i-1}^o = \epsilon_{0i}^o + \epsilon_{Ki}\,\{1 - (t_{i-1}/t_{0i})^{m_i}\}$$

yields:

$$(8.5) \qquad t_{i-1} = t_{0_i}\left[\frac{-\,\epsilon_{i-1}^o + \epsilon_0^o + \epsilon_{Ki}}{\epsilon_K}\right]^{1/m_i}$$

Consequently:

$$t_i = t_{i-1} + \Delta t_i \tag{8.6}$$

Eqs. (4.27) and (4.28) give then the tangent moduli at time t_i and the above computation can be repeated by using the stiffness matrix $[K]_i$ corresponding to average values between the moduli $K_{t,i-1}$, $G_{t,i-1}$, and $K_{t,i}$, $G_{t,i}$. After this iteration or after a repeated iteration we can obtain, by using Eqs. (4.9) and (4.10) and by taking into account the reference time t_{0_i} and the relative time t_i (Eqs. 8.4 to 8.6), the improved values for ϵ_i^o and γ_i^o. With these strains and with the last approximation for octahedral stresses $\sigma_i^{o\prime}$ and τ_i^o the secant moduli

$$K_{si} = 1/3\,(\sigma_i^{o\prime}/\epsilon_i^o)\,, \qquad G_{si} = \tau_i^o/\gamma_i^o \tag{8.7),(8.8}$$

and the corresponding stiffness matrix $[K_s]_i$ can be computed. By applying the finite element procedure in the form

$$\{U\}_i = [K_s]_i^{-1}\,\{P\}_i \tag{8.9}$$

a new, improved approximation for the displacements $\{U\}_i$ is obtained.

The corresponding strain and stress vectors are then given by the relations:

$$\{\epsilon\}_i = [P_s]_i\,\{U\}_i \tag{8.10}$$

$$\{\sigma\}_i = [C_s]_i\,\{\epsilon\}_i \tag{8.11}$$

$[P_s]_i$ and $[C_s]_i$ denoting the strain matrix and the elastic matrix resp., corresponding to the secant moduli. The resulting octahedral stresses $\sigma_i^{o\prime}$ and τ_i^o determine new values of rheological parameters. After having taken into account the creep effect in the same way as before (Eqs. 8.4 to 8.6 and 4.9, 4.10), the computation based on the secant moduli can be repeated. The values of displacements, strains, stresses and strain speeds obtained at the last iteration are then used as starting values for an analogous computation in the following time interval.

With the load increase finished, the initial computation phase with tangent moduli has to be omitted and only iterations with secant moduli can be used.

(b) Creep Effects According to Buisman's Law

When in the rheological equations of the non-linear Kelvin body (4.9) and (4.10) the parameters m and n tend to zero ($m \to 0$, $n \to 0$), the equations have to be replaced by Eqs. (4.48) and (4.49) corresponding to Buisman's logarithmic law of the secondary consolidation:

$$\epsilon^o = \epsilon_0^o + \beta_o \ln(\dot{\epsilon}_0^o / \dot{\epsilon}^o) \tag{4.48}$$

$$\gamma^o = \gamma_0^o + \beta_\gamma \ln(\dot{\gamma}_0^o / \dot{\gamma}^o) \tag{4.49}$$

By an analogous deduction as presented in Section 4, the following expressions for tangent moduli can be obtained (Šuklje, 1977):

$$K_t = K_0 \{ 1 + 3 K_0 B_o \ln(t/t_0) \}^{-1} \tag{8.12}$$

$$G_t = G_0 \{ 1 + G_0 B_\gamma \ln(t/t_0) \}^{-1} \tag{8.13}$$

The parameters K_0, B_o and G_0, B_γ have been defined by Eqs. (4.20), (4.22), (4.24) and (4.26).

The numerical procedure as presented for soils corresponding to the non-linear Kelvin body remains unchanged, with the following simplification of Eqs. (8.4) and (8.5):

$$t_{0_i} = \beta_{0_i} / \dot{\epsilon}_0^o = \beta_{\gamma i} / \dot{\gamma}_0^o \tag{8.14}$$

$$t_{i-1} = t_{0_i} \exp\{ (\epsilon_{i-1}^o - \epsilon_{0_i}^o) / \beta_{0_i} \} \quad \text{or} \tag{8.15-a}$$

$$t_{i-1} = t_{0_i} \exp\{ (\gamma_{i-1}^o - \gamma_{0_i}^o) / \beta_{\gamma i} \} \tag{8.15-b}$$

(c) Creep Increase with a Power of Time

With $\epsilon_K = -\epsilon_0^o$, $\gamma_K = \gamma_0^o$, Eqs. (4.9) and (4.10) get simplified:

$$\epsilon^o = \epsilon_0^o (t/t_0)^m \tag{8.16}$$

$$\gamma^o = \gamma_0^o (t/t_0)^n \tag{8.17}$$

The strain parameters ϵ_0^o , γ_0^o, β_o and β_γ can be found by presenting the test data in the form of charts

$$\log \epsilon^o = \log \epsilon_0^o + m \log(t/t_0) \tag{8.16-a}$$

$$\log \gamma^o = \log \gamma_0^o + n \log(t/t_0) \tag{8.17-a}$$

The obtained parameter values are used for the construction of four families of plots according to Eqs. (4.11), (4.13), (4.14) and (4.16). Alternatively, according to Eqs. (4.36) and (4.37), equations (8.16) and (8.17) can be written in terms of strain-speed ratios:

$$\epsilon^o = \epsilon_0^o (\dot{\epsilon}^o / \dot{\epsilon}_0^o)^\alpha \tag{8.18}$$

$$\gamma^o = \gamma_0^o (\dot{\gamma}^o / \dot{\gamma}_0^o)^\beta \tag{8.19}$$

whereby

$$\alpha = m/(m-1) \quad , \qquad \beta = n/(n-1) \qquad (4.38)\ ,(4.39)$$

With $\epsilon_K = -\epsilon_0^o$ and $E_K = K_0$, expressions (4.27) and (4.28) for the tangent moduli reduce to:

$$K_t = K_0 \{ (t/t_0)^m [1 + 3 K_0 B_0 \epsilon_0^o \ln(t/t_0)] \}^{-1} \tag{8.20}$$

(8.21) $$G_t = G_0 \{ (t/t_0)^n [1 + G_0 B_\gamma \gamma_0^o \ln(t/t_0)] \}^{-1}$$

The procedure as presented for the non-linear Kelvin body remains again unchanged with the following modifications of Eqs. (8.4) and (8.5):

(8.22) $$t_{0_i} = - m_i (\epsilon_{0_i}^o / \dot{\epsilon}_0^o)$$

(8.23) $$t_{i-1} = t_{0_i} (\epsilon_{i-1} / \epsilon_{0_i}^o)^{1/m_i}$$

Furthermore, Eqs. (4.9) and (4.10) have to be replaced by Eqs. (8.16) and (8.17).

Examples

(a) Drained Conditions

For the non-linear rheological relationships corresponding to Buisman's law of the secondary consolidation a computer program has been prepared by B. Majes (see Šuklje and Majes, 1975, Šuklje, 1977) enabling the analysis of stresses and strains in a layered half-plane with different rheological relationships for single layers and different relationships $k = k(e)$ between the coefficient of permeability and the void ratio. The algorithm of the program follows the above presented computation scheme.

Figs. 8.1 to 8.3 present some results of the application of this program to a single layer of limited thickness subjected to a symmetrical triangular load strip. The layer consists of the highly compressible silty clay whose drained rheological relationships have been presented in Figs. 4.2 to 4.5. The load was applied with the speed $q = 0.96$ kp cm^{-2} year^{-1} = 94.14 kPa year^{-1} . The boundary conditions and the network of finite elements are shown in Fig. 8.3. Fig. 8.1 presents the comparison of settlements (v) during the load increase with the values obtained if disregarding viscous effects. The increase of settlements in the load axis after the end of the load increase is shown in Fig. 8.2, and the corresponding yielded elements in Fig. 8.3.

The finite elements have been considered to yield if the Mohr stress circle reached the strength envelope. When approaching failure, shear strains tend to infinity (cf. Fig. 4.4) and the Poisson ratio to 0.50. In order to enable the numerical application of the step by step elastic analysis, the zero value of the tangent shear modulus has been replaced by a very small value ($G = 0.0033$ kp cm^{-2} in our case) and the Poisson ratio $\nu = 0.50$ by a somewhat smaller value, close to 0.50 ($\nu = 0.499$ in our case).

After the end of the load increase, the creep, at $q = 0.48$ kp cm^{-2}, develops faster than with the logarithm of time. Such an intensive creep can be interpreted by the progressive yielding of elements below the load strip. (Further results and comments have been presented by Šuklje, 1977.)

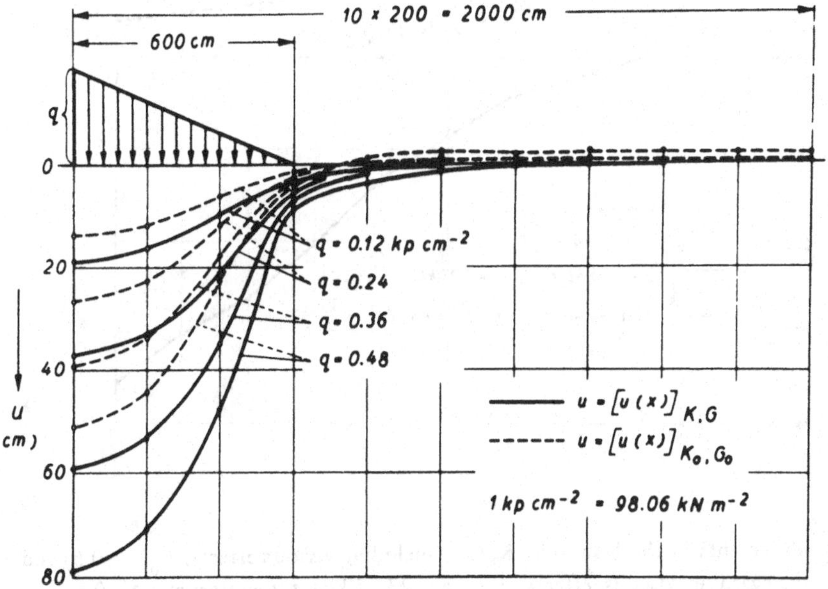

Fig. 8.1 - Settlements during the load increase ($\dot{q}$ = 94.14 kPa year^{-1}): K,G .. considering viscous effects, K_0, G_0 .. disregarding viscous effects.

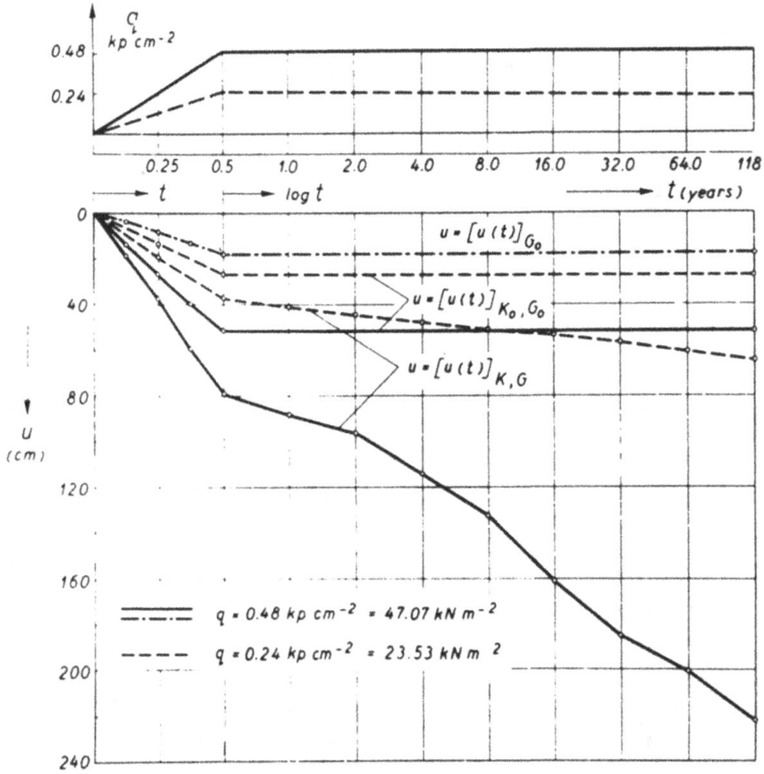

Fig. 8.2 - Settlements in the load axis: K, G . . including viscous effects, $G_0(\nu \rightarrow 0.5)$ and K_0, G_0 . . disregarding viscous effects. For q = 23.53 kPa t denotes: at t $\leqslant$ 0.5 year double value of the real time, at t $\geqslant$ 0.5 year the real time + 0.25 year.

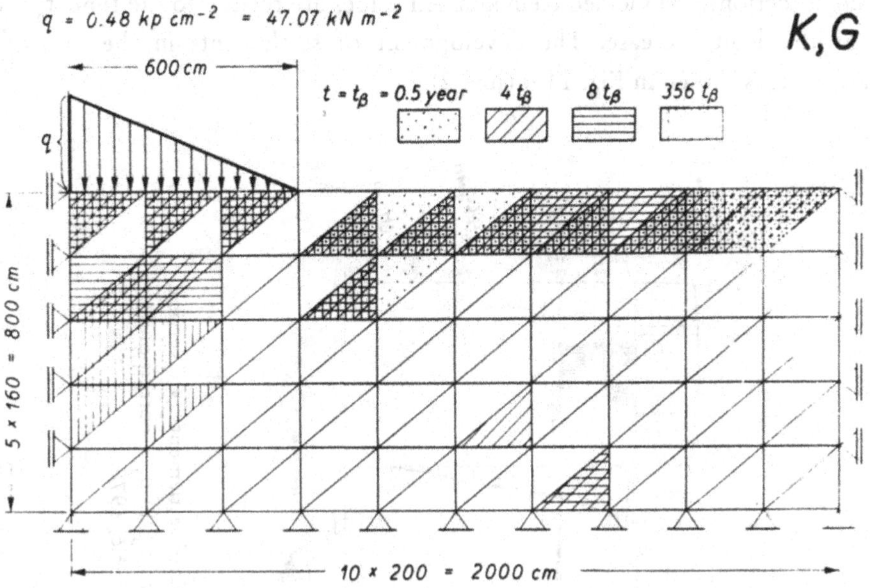

Fig. 8.3 - Yielding at the end (t_β) and after the end of the load increase.

(b) Undrained Conditions

When the previously explained numerical procedure is applied for the computation of pure deviatoric displacements prevailing in the early consolidation phase of poorly permeable soils, the rheological relationships as obtained from undrained saturated samples and expressed in terms of total stresses may be used. The compression modulus becomes infinite and the Poisson ratio equal to 0.50; in the numerical computation a somewhat smaller value, close to 0.5, has to be taken into account.

Some results of the application of the procedure to a test embankment on a multi-layer soil with two compressible upper layers (the undrained rheological relationships for the upper silty layer have been presented in Figs. 4.6 to 4.9) are shown in Fig. 8.4 (reproduced after Šuklje and Kovačič, 1978). The figure presents: (1) the network of finite elements and the boundary conditions considered in the analysis, (2) surface settlements (v) and three vertical diagrams of vertical displacements (v), (3) vertical pressure increments along one horizontal and two vertical cross-sections, (4) yielded elements. All values are related to the time t_β at the end of the load increase. The development of settlements in the axis of the embankment is shown in Fig. 11.4 (line 2).

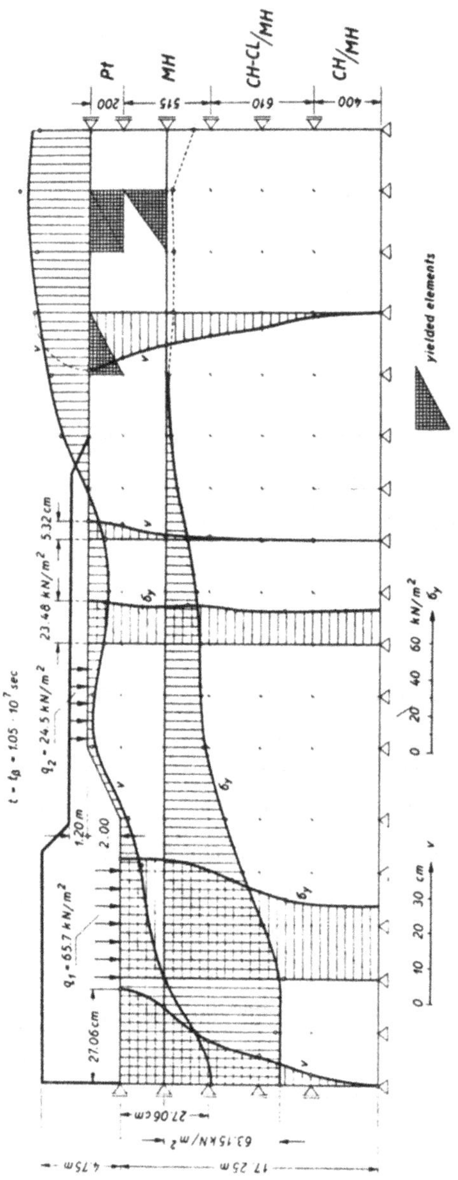

Fig. 8.4 - Some results of a finite element analysis of undrained settlements: v = vertical displacement components at the end of the load increase (t_β), σ_y = corresponding vertical pressure increments. Reproduced after Šuklje and Kovačič, 1978.

9. DIFFUSION EQUATION

Deduction of the General Form of the Differential Equation of the One-Dimensional Consolidation

The differential equation of the one-dimensional consolidation of soils was developed by K. Terzaghi (1923). The equation was based on the following assumptions: the voids of the soil are completely filled with water; both the water and the solid constituents of the soil are perfectly incompressible; Darcy's law (in the form of Equation $v = k\ i$) is strictly valid; the coefficient of permeability k is a constant; the time lag of consolidation is due entirely to the low permeability of the soil; the soil is laterally confined; both, the total and the effective normal stresses are the same for every point on any horizontal section through the soil and for every stage of the process of consolidation; the void ratio decrease is linearly proportional to the effective stress increase.

Terzaghi's classical solution can be treated as a special case of a more general form of the diffusion equation allowing for partly saturated soils, for a modified form of Darcy's law (Equation 9.4-a), for the variation of the coefficient of permeability with void ratio, for the viscous properties of soil and for non-linear stress-strain relationships. This generalized form of the consolidation equation will be presented by making use of Florin's deduction (Florin 1961, for saturated soils: 1938). In a somewhat restricted form it was firstly deduced by Biot (1935, 1941).

In this lecture a summarized deduction of the equation of consolidation will be given (for details see e.g. Šuklje, 1969-a). The following notation will be used:

$n = e/(1 + e)$ porosity, $n = n_a + n_w$ (9.1)

n_a , n_w porosity related only to voids filled with air and water respectively

$e_a = n_a/(1 - n)$, $e_w = n_w/(1 - n)$ the corresponding void ratios (9.2)

v , $\mathring{v}$, w the respective speeds of the liquid, solid and gaseous constituent of the soil related to the total cross-section area of the soil

$v_e = v/n_w$, $\mathring{v}_e = \mathring{v}/(1 - n)$ the respective effective speeds of the liquid and solid constituent of the soil (9.3)

ρ	the density of the gas (air)
M	Henry's coefficient of the solubility of the air in the water
H	hydraulic potential height (piezometric level)
H_o	H corresponding to the limit hydraulic gradient $i_o = \partial H_o/\partial z$

(9.4) $\bar{H} = H - H_o\,, \qquad \dfrac{\partial \bar{H}}{\partial z} = \dfrac{\partial H}{\partial z} - \dfrac{\partial H_o}{\partial z} = i - i_o$

according to the general form of Darcy's law (see Florin, 1959):

(9.4-a)
$$v = k(i - i_o) \quad \text{when} \quad i \geqslant i_o$$
$$v = 0 \qquad\qquad \text{when} \quad i < i_o$$

$u_w^*\,,\ u_a^*$	pore-water and pore-air pressure respectively
$u_{wo}^*\,,\ u_{ao}^*$	their initial values (at $t = 0$)
$u_w\,,\ u_a$	pore-water and pore-air excess pressure resp.:

(9.5)
$$u_w = u_w^* - u_w^o\,, \quad u_a = u_a^* - u_a^o$$

$u_w^o\,,\ u_a^o$	pore-pressure component remaining constant during consolidation and not influencing the consolidation process; the atmospheric pressure is included.

Basic relations are:

(a) continuity condition for the liquid phase of the soil:

(9.6)
$$\frac{\partial v}{\partial z} + \frac{\partial n_w}{\partial t} = 0$$

(b) continuity condition for the solid phase of the soil:

$$\frac{\partial \dot{v}}{\partial z} + \frac{\partial(1-n)}{\partial t} = 0 \qquad (9.7)$$

(c) continuity condition for the gaseous phase of the soil:

$$\frac{\partial(n_a \rho)}{\partial t} + M\frac{\partial(\rho n_w)}{\partial t} + \frac{\partial(\rho w)}{\partial z} = 0$$

yielding, if neglecting the relatively small values $M\frac{\partial n_w}{\partial t}\rho$ and $w\frac{\partial \rho}{\partial z}$,

$$\frac{\partial n_a}{\partial t} + (n_a + Mn_w)\frac{1}{\rho}\frac{\partial \rho}{\partial t} + \frac{\partial w}{\partial z} = 0 \qquad (9.8)$$

In the above continuity condition (c) Henry's law has been taken into account according to which the mass of the air dissolved in the water is linearly proportional to the density of the air and to the quantity of the dissolving water.

(d) Darcy's law which may be written in the form:

$$v_e - \dot{v}_e = -\frac{k}{n_w}\frac{\partial \bar{H}}{\partial z}$$

which yields, after substituting the expressions (9.3), differentiating with respect to to z and neglecting the relatively small term $\dot{v}\,\partial/\partial z\,[(n_w/(1-n)]$,

$$\frac{\partial v}{\partial z} = \frac{n_w}{1-n}\frac{\partial \dot{v}}{\partial z} - \frac{\partial}{\partial z}\left(k\frac{\partial \bar{H}}{\partial z}\right)$$

(e) The gas state equation

$$\rho = \frac{1}{rT}\,p$$

r being the specific gas constant, T the absolute temperature and p the gas pressure.

In the case of an isothermal change of state we have (see Eq. 9.5)

$$\frac{1}{\rho}\frac{\partial \rho}{\partial t} = \frac{1}{p}\frac{\partial p}{\partial t} = \frac{1}{u_a^*}\frac{\partial u_a^*}{\partial t} = \frac{1}{u_a^o + u_a}\frac{\partial u_a}{\partial t} \tag{9.10}$$

(f) The assumption is taken that the air bubbles cannot move between the grains and that, as a result, the average values of the real velocities of the grains and of the air bubbles are equal throughout the consolidation process:

$$\mathring{v}_e = w_e$$

Thus

$$\mathring{v}\,/\,(1-n) = w/n_a \;\rightarrow\; w = n_a\,\mathring{v} : (1-n) \tag{9.11}$$

By adding the continuity conditions for the liquid, solid and gaseous phases (Eqs. 9.6, 9.7, 9.8), we obtain the relation

$$\frac{\partial}{\partial z}(v + \mathring{v} + w) + (n_a + Mn_w)\,\frac{1}{\rho}\frac{\partial \rho}{\partial t} = 0 \tag{9.12}$$

After having inserted relations (9.9) and (9.11), neglected the relatively small term $\mathring{v}\,\partial e_w/\partial z$ and considered equations (9.7) and (9.10), the above equation yields:

$$\frac{\partial e}{\partial t} + \frac{e_a + Me_w}{u_a^o + u_a}\frac{\partial u_a}{\partial t} = (1 + e)\,\frac{\partial}{\partial z}\left(k\,\frac{\partial \bar{H}}{\partial z}\right) \tag{9.13}$$

This is the general form of the diffusion equation for partly saturated soils provided that the air is considered static, trapped by the grain skeleton.

Some Alterations and Specifications of the Diffusion Equation for the One-Dimensional Consolidation

The piezometric height H equals to

$$H = z + (u_w^* / \gamma_w) \tag{9.14}$$

The total pore-water pressure u_w^* can be considered to be composed of the following three parts:

$$u_w^* = u_w^o + u_w + u_{wi} \tag{9.15}$$

u_{wi} denoting the part of the pore-pressure corresponding to the limit hydraulic gradient i_0 :

$$i_0 = \frac{\partial u_{wi}}{\partial z} \frac{1}{\gamma_w} \tag{9.16}$$

Consequently:

$$\frac{\partial \bar{H}}{\partial z} = \frac{\partial H}{\partial z} - i_0 = 1 + \frac{1}{\gamma_w} \left(\frac{\partial u_w^o}{\partial z} + \frac{\partial u_w}{\partial z} \right) \tag{9.17}$$

The terms

$$\left(1 + \frac{1}{\gamma_w} \frac{\partial u_w^o}{\partial z}\right)$$

represent the gradient of the initial stationary seepage not influencing the consolidation process. Our further deductions are restrained to hydrostatic fields where

(9.18)
$$1 + \frac{1}{\gamma_w} \frac{\partial u_w^o}{\partial z} = 0$$

Then we can write

(9.19)
$$\frac{\partial}{\partial z} \left(k \frac{\partial \bar{H}}{\partial z} \right) = \frac{k}{\gamma_w} \frac{\partial^2 u_w}{\partial z^2} + \frac{1}{\gamma_w} \frac{\partial k}{\partial z} \frac{\partial u_w}{\partial z}$$

and the diffusion equation gets the form:

(9.20)
$$\frac{\partial e}{\partial t} + \frac{e_a + M e_w}{u_a^0 + u_a} \frac{\partial u_a}{\partial t} = \frac{1}{\gamma_w} (1 + e) \frac{\partial}{\partial z} \left(k \frac{\partial u_w}{\partial z} \right)$$

The difference between the pore-air (u_a^*) and pore-water pressure (u_w^*) is due to the surface tension of the pore-water (u_m, expressed as absolute value):

(9.21)
$$u_a^* = u_w^* + u_m$$

With the simplifying assumption that the surface tension u_m remains constant during the entire consolidation process

(9.22)
$$u_m = \text{const}$$

and with the notation.

(9.23)
$$u_w \equiv u \, , \quad u_w^* \equiv u^*$$

Eq. (9.20) becomes simplified:

(9.24)
$$\frac{\partial e}{\partial t} + \frac{e_a + M e_w}{u^0 + u_m + u} \frac{\partial u}{\partial t} = \frac{1 + e}{\gamma_w} \frac{\partial}{\partial z} \left(k \frac{\partial u}{\partial z} \right)$$

or, by substituting

(9.25)
$$\frac{e_a + M e_w}{u^0 + u_m + u} = \beta$$

$$\frac{\partial e}{\partial t} + \beta \frac{\partial u}{\partial t} = \frac{1 + e}{\gamma_w} \frac{\partial}{\partial z} \left(k \frac{\partial u}{\partial z}\right) \qquad (9.26)$$

When solving the above equation, the principle of effective stress has to be considered in the form

$$\sigma' = \sigma - u_n^* \qquad (9.27)$$

According to Eq. (1.1) and taking into account relations (9.21) and (9.23) we have

$$\sigma' = \sigma - \{ (\xi u^* + (1 - \xi)(u^* + u_m) \}$$ yielding:

$$\sigma' = \sigma - (1 - \xi) u_m - u^* \qquad (9.28)$$

Inserting, for $u^* \equiv u_w^*$, Eq. (9.15) we obtain:

$$\sigma' = \sigma - \{ (1 - \xi) u_m + u^0 + u_i \} - u$$

or, with the substitution

$$\sigma - \{ (1 - \xi) u_m + u^0 + u_i \} = \sigma_r \qquad (9.29\text{-a})$$

$$\sigma' = \sigma_r - u \qquad (9.29\text{-b})$$

Since the limit hydraulic gradient depends on the void ratio:

$$i_0 = i_0(e) \qquad (9.30)$$

u_i in Eq. (9.29-a) can be obtained by the integration:

(9.31) $$u_i = \int_{z=z_0}^{z} \frac{\partial u_i}{\partial z} dz = \gamma_w \int_{z=z_0}^{z} i_0(e)\ dz$$

$$i_0(e) = i_0\{e(z)\} = i_0(z)$$

For saturated soils $\beta = 0$. Assuming $k = \text{const}$, Eq. (9.26) reduces to

(9.32) $$\frac{\partial e}{\partial t} = \frac{(1 + e)\ k}{\gamma_w} \frac{\partial^2 u}{\partial z^2}$$

or

(9.33) $$-\frac{\partial \epsilon_v}{\partial t} = \frac{k}{\gamma_w} \frac{\partial^2 u}{\partial z^2}$$

For linear stress-strain conditions

(9.34) $$\frac{\partial \sigma'}{\partial \epsilon_v} = M_v$$

M_v denoting the compression modulus of oedometer tests. Considering the principle of effective stress according to Eq. (9.29)

(9.35) $$u = \sigma_r - \sigma'$$

Eq. (9.33) can be written in the form

(9.36) $$\frac{\partial u}{\partial t} = \frac{\partial \sigma_r}{\partial t} + \frac{k\,M_v}{\gamma_w} \frac{\partial^2 u}{\partial z^2}$$

Denoting

$$k\,M_v/\gamma_w = c_v = \text{coefficient of consolidation} \tag{9.37}$$

and assuming a sudden load increase ($\partial\sigma_r/\partial t = 0$) and zero limit hydraulic gradient ($i_0 = 0$), we get the classical form of the differential equation of one-dimensional consolidation (Terzaghi, 1923):

$$\frac{\partial u}{\partial t} = c_v \frac{\partial^2 u}{\partial z^2} \tag{9.38}$$

Diffusion Equation of the Three-Dimensional Consolidation in Cartesian Coordinates

If the drainage is allowed to occur in all directions, the continuity conditions for the liquid, solid and gaseous phases respectively, take the following forms:

$$\frac{\partial v_x}{\partial x} + \frac{\partial v_y}{\partial y} + \frac{\partial v_z}{\partial z} + \frac{\partial n_w}{\partial t} = 0 \tag{9.39}$$

$$\frac{\partial \bar{v}_x}{\partial x} + \frac{\partial v_y}{\partial y} + \frac{\partial v_z}{\partial z} + \frac{\partial (1-n)}{\partial t} = 0 \tag{9.40}$$

$$\frac{\partial w_x}{\partial x} + \frac{\partial w_y}{\partial y} + \frac{\partial w_z}{\partial z} + \frac{\partial n_a}{\partial t} + \frac{n_a + M n_w}{u_a^0 + u_a} \cdot \frac{\partial u_a}{\partial t} = 0 \tag{9.41}$$

In a way analogous to that in which the general equation of one-dimensional consolidation (9.13) was deduced in the foregoing Section, the generalized continuity conditions (9.29) to (9.31) lead to the following general equation of three-dimensional consolidation:

$$\frac{\partial e}{\partial t} + \frac{e_a + Me_w}{u_a^0 + u_a} \cdot \frac{\partial u_a}{\partial t} = (1 + e)\left[\frac{\partial}{\partial x}\left(k_x \frac{\partial \bar{H}_x}{\partial x}\right) + \frac{\partial}{\partial y}\left(k_y \frac{\partial \bar{H}}{\partial y}\right) + \frac{\partial}{\partial z}\left(k_z \frac{\partial \bar{H}_z}{\partial z}\right)\right] \tag{9.42}$$

whereby:

$$\frac{\partial \bar{H}_x}{\partial x} = \frac{\partial H}{\partial x} - i_{ox} \tag{9.43}$$

and similarly for the directions y and z.

When the initial pore-water potential field is hydrostatic and the limit gradients i_{ox}, i_{oy}, i_{oz} are zero, and when, furthermore, the surface tension u_m is assumed to remain constant during the consolidation, Equation (9.32) gets simplified:

$$\frac{\partial e}{\partial t} + \beta \frac{\partial u}{\partial t} = \frac{1 + e}{\gamma_w}\left[\frac{\partial}{\partial x}\left(k_x \frac{\partial u}{\partial x}\right) + \frac{\partial}{\partial y}\left(k_y \frac{\partial u}{\partial y}\right) + \frac{\partial}{\partial z}\left(k_z \frac{\partial u}{\partial z}\right)\right] \tag{9.44}$$

with β according to Equation (9.20). The principle of effective stress has to be adapted to spheric stresses in the form (see Eqs. 9.29-a,b):

$$\sigma'_{oct} = \sigma_{oct_r} - u \tag{9.45}$$

$$\sigma_{oct_r} = \sigma_{oct} - \{(1 - \chi)u_m + u^0\} \tag{9.46}$$

For isothropic homogeneous soils

$$k_x = k_y = k_z = k = \text{const} \tag{9.47}$$

Equation (9.44) yields:

$$\frac{\partial e}{\partial t} + \beta \frac{\partial u}{\partial t} = \frac{(1 + e)k}{\gamma_w} \Delta u \tag{9.48}$$

$$\Delta = \frac{\partial}{\partial x^2} + \frac{\partial}{\partial y^2} + \frac{\partial}{\partial z^2}$$

For saturated soils ($\beta = 0$), the above equation can be written also in the form:

$$-\frac{\partial \epsilon_v}{\partial t} = \frac{k}{\gamma_w} \Delta u \tag{9.49}$$

Consolidation of Saturated Elastic Media

The seepage forces γ_w i due to the consolidation appear, in the equilibrium conditions for elastic media, as volume forces:

$$\Delta\xi + \frac{1}{1-2\nu}\frac{\partial \epsilon_v}{\partial x} - \frac{1}{G}\frac{\partial u}{\partial x} = 0 \tag{9.50-a}$$

$$\Delta\eta + \frac{1}{1-2\nu}\frac{\partial \epsilon_v}{\partial y} - \frac{1}{G}\frac{\partial u}{\partial y} = 0 \tag{9.50-b}$$

$$\Delta\zeta + \frac{1}{1-2\nu}\frac{\partial \epsilon_v}{\partial z} - \frac{1}{G}\frac{\partial u}{\partial z} = 0 \tag{9.50-c}$$

where ξ, η, ζ are the displacements in the x, y, z directions respectively of a Cartesian system of coordinates, ϵ_v the volume strain:

$$\epsilon_v = \epsilon_x + \epsilon_y + \epsilon_z = \frac{\partial \xi}{\partial x} + \frac{\partial \eta}{\partial y} + \frac{\partial \zeta}{\partial z}$$

By differentiating the first of the equations (9.38) with respect to x, the second with respect to y and the third with respect to z, by adding the equations obtained, and by reducing the resulting expression we obtain:

(9.51) $$\frac{2(1-\nu)}{1-2\nu} G \Delta\epsilon_v = \Delta u$$

Expressing ν and G by Lamé's coefficients λ and μ

(9.52), (9.53) $$\lambda = K - \frac{2}{3} G , \qquad \mu \equiv G$$

Equation (9.51) obtains the form

(9.54) $$(\lambda + 2\mu) \Delta\epsilon_v = \Delta u$$

Inserting the above expression into the equation of consolidation (9.49) we get:

(9.55) $$-\frac{\partial \epsilon_v}{\partial t} = c_v \Delta\epsilon_v$$

where the coefficient of consolidation c_v is defined by

(9.56) $$c_v = \frac{k(\lambda + 2\mu)}{\gamma_w}$$

Equation (9.41) was first deduced by Biot (1935 and 1941).

Diffusion Equation in Cylindrical Coordinates

In cylindrical coordinates, Equation (9.44) gets the form:

$$\frac{\partial e}{\partial t} + \frac{e_a + M e_w}{u^0 + u_m + u} \cdot \frac{\partial u}{\partial t} = \frac{1+e}{\gamma_w} [\frac{\partial}{\partial r}(k_r \frac{\partial u}{\partial r}) + \frac{k_r}{r} \cdot \frac{\partial u}{\partial r} + \frac{\partial}{\partial z}(k_z \frac{\partial u}{\partial z})]$$

(9.57)

with the following meaning of the new symbols:

k_r , k_z coefficients of permeability in the directions r and z respectively

r , z radial and axial coordinate respectively

When the coefficients k_r and k_z are constant and their ratio is denoted κ :

$$\kappa = k_r/k_z \, , \quad k_r = \text{const}_r \, , \quad k_z = \text{const}_z \equiv k \, ,$$

Equation (7.82) gets simplified. For saturated soils ($\beta = 0$) we have:

$$\frac{\partial e}{\partial t} = \frac{(1+e)k}{\gamma_w} \left\{ \kappa \left(\frac{\partial^2 u}{\partial r^2} + \frac{1}{r} \frac{\partial u}{\partial r} \right) + \frac{\partial^2 u}{\partial z^2} \right\} \tag{9.58}$$

10. NUMERICAL ANALYSIS OF THE INFLUENCE OF PARAMETERS GOVERNING THE CONSOLIDATION OF VISCOUS SOILS

One-Dimensional Consolidation of Partly Saturated Viscous Soils

(a) Deduction of the System of Fundamental Equations Suitable for Numerical Computation

The differential equation of consolidation will be taken into account in the form (9.19) corresponding to the assumption $u_m = \text{const}$. In order to avoid unnecessary complications in analysing the influence of the saturation degree, the initial density (effect of the previous secondary consolidation), of the load interval and of the layer thickness (seepage path), a constant coefficient of permeability k equal to the average quantity appearing in the treated stress interval, will be considered. With this assumption and with the substitution

$$u^o + u_m = u_a^o \tag{10.1}$$

Equation (9.19) can be written in the form:

$$\frac{\partial e}{\partial t} = \frac{(1 + e)}{\gamma_w} \frac{\partial^2 u}{\partial z^2} - \frac{e_a + Me_w}{u_a^o + u} \cdot \frac{\partial u_a}{\partial t} \tag{10.2}$$

(For notation see Section 9.) The speed of the void ratio change $\partial e/\partial t$ can be expressed from the experimentally obtained rheological relationship $R(e, \dot{e}, \sigma') = 0$ in the form

$$\frac{\partial e}{\partial t} = c(e, \sigma') \tag{10.3}$$

$\sigma' = \sigma'(z, t)$ is the effective stress in the seepage direction.

As the occluded air bubbles have been assumed to be trapped by the grain skeleton, so that the air quantity does not change during the consolidation, Boyle's law can be written as follows:

$$e_a u_a^* = (e_{ao} - \Delta e_{ao}) u_{ao}^* \tag{10.4}$$

$$\Delta e_{ao} = e_o S_{ro} (u - u_o) \frac{M}{u_{ao}^*} < e_{ao} \tag{10.5}$$

The suffix "o" designates the value at the time $t = 0$:

$$(10.6) \qquad e_o = e(z,0)\,, \quad e_{ao} = e_a(z,0)\,, \quad u^*_{ao} = u^*_a(z,0)\,, \quad u_o = u(z,0)$$

Δe_{ao} denotes the change of the air-filled void ratio due to the solution of the air in the water at the pore-water excess pressure increase from u_o to u

S_{ro} is the saturation degree at time $t = 0$: $S_{ro} = e_{wo}/e_o$

The total pressure u^*_a equals

$$(10.7) \qquad u^*_a = u + u^o_a$$

where u^o_a (see Equ. 10.1) denotes the pore-air pressure component due to the stationary part of the hydraulic potential field, the atmospheric pressure and the surface tension included, i.e. the component that remains constant during the consolidation. Similarly we have:

$$(10.8) \qquad u^*_{ao} = u_o + u^o_a$$

Taking into account the relations (10.7) and (10.8), Equations (10.4) and (10.5) yield

$$(10.9) \qquad e_a = -\,e_o\,S_{ro}\,M + \frac{u_o + u^o_a}{u + u^o_a}\,[e_{ao} + e_o\,S_{ro}\,M]$$

When the condition (10.5) $\Delta e_{ao} < e_{ao}$ is not fulfilled, Equation (10.9) has to be replaced by

$$(10.10) \qquad e_a = 0$$

According to the principle of effective stresses (1.1) we have

$$(10.11) \qquad \sigma' = \sigma^* - [u^*_a - \chi\,(u^*_a - u^*_w)]\,, \qquad 0 \leqslant \chi \leqslant 1$$

Taking into account the relation (9.21), the assumption (9.22) as well as the substitutions (9.23) and the following one:

$$\sigma^* - u^\circ = \sigma \tag{10.12}$$

Equation (10.11) yields:

$$\sigma' = \sigma - [u + (1 - \chi)u_m] \tag{10.13}$$

According to the experimental data of Bishop and Donald (1961) we shall take the rough approximation:

$$\chi \cong S_r \cong S_{ro} \tag{10.14}$$

In this way, Equation (10.13) gets further simplified:

$$\sigma' = \sigma - (1 - S_{ro})\,u_m - u \tag{10.15}$$

or, with the substitution

$$\sigma - (1 - S_{ro})\,u_m = \sigma_r \;, \tag{10.16}$$

$$\sigma' = \sigma_r - u \tag{10.15-a}$$

For the purpose of our analysis, the above equation should be rewritten in the form

$$\sigma_r(z, t) = \sigma'(z, t) + u(z, t) \tag{10.17}$$

In the statically determinate stress conditions the reduced total stresses σ_r, acting in the seepage direction, are known at any time:

$$\sigma_r(z, t) = \sigma_r(z, 0) + q(t) \tag{10.18}$$

The function q(t) is given.

Taking into account the connections (10.3), (10.9) and (10.10) respectively, (10.17) and (10.18), the system of equations governing the consolidation process can be written as follows:

$$c(e, \sigma') = -a(e)\frac{\partial^2 u}{\partial z^2} + b(e, u, z)\frac{\partial u}{\partial t} \tag{10.19}$$

$$\frac{\partial e}{\partial t} = c(e, \sigma') \tag{10.3}$$

(b) Boundary and Initial Conditions

The paper "Consolidation of partly saturated viscous soils" by Šuklje and Kozak (1974) presents the numerical solution of the differential equation of consolidation for viscous soils of non-linear deformability and partial saturation and shows the influence of various factors onto the consolidation process for the case of a horizontal layer of uniform thickness (Fig. 10.1). The layer was assumed to consist of a normally consolidated lacustrine clay (from the Lake of Skadar, see Šuklje, 1957, and Šuklje – Simončič, 1972) whose deformability was proved to be governed by the rheological equation of the type (1.11):

$$\frac{\partial e}{\partial t} = c(e, \sigma') = \dot{e}_o \exp \frac{A + B \ln \frac{\sigma'}{\sigma_o} - e}{C + D \ln \frac{\sigma'}{\sigma_o}} \tag{10.20}$$

The upper and lower boundary of the layer were assumed to be pervious:

$$u(0, t) = u(h, t) = 0 \tag{10.21}$$

The initial vertical total pressure σ_{oz} at the upper boundary was supposed to be given. The initial total pressure $\sigma(z, 0)$ increases with the depth according to the gravity field of partly saturated soils below the ground-water level. During the entire consolidation process the term $(1 - S_{ro})\, u_m$ in Equation (10.16) was supposed to be included in the pressure σ_{oz} and to be constant throughout the layer.

Thus, the function $\sigma_p(z, 0)$ had to satisfy the differential equation

$$\frac{d\sigma_r(z,0)}{dz} = \frac{d\sigma(z,0)}{dz} = \frac{(\gamma_s - \gamma_w) - \gamma_w \; e(z,0)\,(1 - S_{ro})}{1 + e(z,0)} \qquad (10.22)$$

at the initial condition:

$$\sigma_r(0, 0) = \sigma_{oz}$$

(γ_s = specific gravity of grains). If on intact samples taken in several depths z the void ratio e(z, 0) and the saturation degree were determined, the total pressure versus depth plot can be ascertained. For the purpose of the above mentioned study

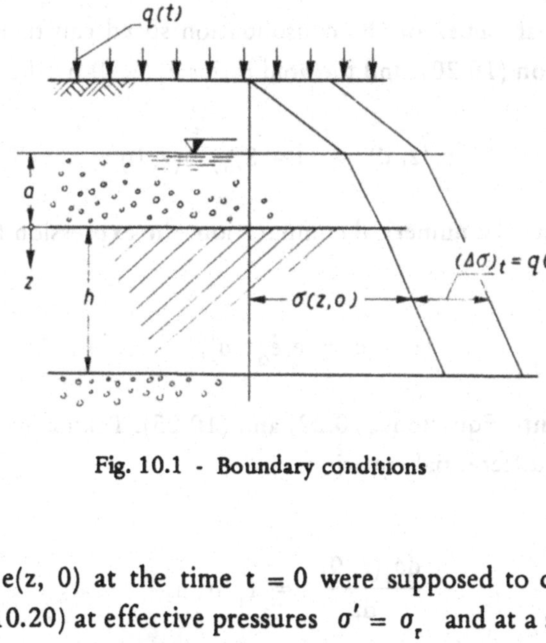

Fig. 10.1 - Boundary conditions

the void ratios e = e(z, 0) at the time t = 0 were supposed to correspond to the isotache equation (10.20) at effective pressures $\sigma' = \sigma_r$ and at a small initial speed $\dot{e}_\alpha$ occurring at the upper boundary of the layer (z = 0) in a late phase of the previous secondary consolidation. The void ratio e can then be obtained from the equation

$$\dot{e}_\alpha = c(e, \sigma') \qquad (10.23)$$

The stationary component u_a^o of total pore-air pressures appearing in Equation (10.2) is, in the treated case (Fig. 10.1), given by the equation

$$(10.24) \qquad u_a^o = p_o + \gamma_w (a + z) \ ,$$

where p_o denotes the atmospheric pressure at the ground-water level. The initial excess pore-pressures u (Equation 10.7) are due to the long-term consolidation at the initial total stress-state and change very little with time. Thus, with $\partial u / \partial t = 0$ Equation (10.19) yields

$$(10.25) \qquad c(e, \sigma') = - a(e) \frac{\partial^2 u}{\partial z^2} \ , \qquad u(0) = u(h) = 0$$

Equations (10.22), (10.23) and (10.25) determine the initial values of u, σ and e. The corrected initial values of the consolidation speed can then be obtained from the isotache equation (10.20), and the void ratios $e_a(z, 0)$ by the relation

$$(10.26) \qquad e_a(z, 0) = (1 - S_{ro})\ e(z, 0)$$

In order to facilitate the numerical computation, the expression for void ratios from Equation

$$(10.27) \qquad e = e(\dot{e}_\alpha , \sigma')$$

has to be inserted into Equations (10.22) and (10.25). Taking into account Equation (10.16) we get the differential equations

$$(10.28) \qquad \frac{d\sigma_r(z, 0)}{dz} = f_1(\sigma, u)$$

$$(10.29) \qquad - \frac{\partial^2 u}{\partial z^2} = f_2(\sigma, u)$$

The system of equations (10.28) and (10.29) defines the initial values of u and σ_r .

(c) Discretization of Differential Equation and Numerical Procedure

The system of the fundamental differential equations (10.19) and (10.3) as well as of the differential equations (10.28) and (10.29) defining the initial conditions were solved by a numerical procedure. The way of discretization and of the successive numerical computation was presented in the paper by Šuklje and Kozak, 1974, and will not be reviewed in this lecture.

The authors have discussed also the computation possibilities and limitations when the dependence of the permeability on the void ratio is taken into account and when the procedure is applied to two-dimensional problems.

(d) Some Results of Numerical Analysis of Factors Governing the Consolidation

By a concrete numerical analysis the influence of the following factors governing the consolidation has been brought to light: the saturation degree (three variations), the degree of the previous secondary consolidation (initial consolidation speed $\dot{e}_\alpha$, three variations), the magnitude of the load interval (two alternations), the thickness of the layer (two alternations). The values of the parameters A, B, C, D, σ_o and $\dot{e}_o$ appearing in the rheological equation (10.20), as well as the total-stress and pore-pressure boundary conditions are shown in Figs. 10.2 and 10.3. The additional load appearing in Equation (10.18) has been assumed to be applied according to the equations:

$$q(t) = q \cdot (t/t_\beta) \qquad 0 \leqslant t \leqslant t_\beta \qquad (10.30\text{-a})$$

$$q(t) = \Delta\sigma = \sigma_\beta - \sigma_\alpha = q \qquad t > t_\beta \qquad (10.30\text{-b})$$

whereby t_β alternates according to the relation

$$t_\beta = 10^6 \, (h/h_o)^2 \text{ sec} , \quad h_o = 525 \text{ cm} \qquad (10.31)$$

Some instructive results of the numerical computations (made with the computer CDC CYBER 72) are shown in Figs. 10.2 to 10.7. (Additional results have been presented in the paper by Šuklje and Kozak, 1974). Figs. 10.2 to 10.5 are related to the computation cases with the layer thickness of 525 cm, while Figs. 10.6 and 10.7 show the influence of different layer thickness: h = 525 and 2.1 cm respectively.

From the above mentioned example the following conclusions can be deduced:

(a) At the same t/h^2 value (t = time after the beginning of the load application, h = thickness of the layer), the excess pore-pressure versus the total pressure increase ratio has been proved to be the smaller, the thinner the layer and the smaller the saturation degree, provided that the initial porosity e_α and the load interval q are the same and the time t_β of the load application proportional to h^2. The excess pore-pressures have been found to increase beyond the time t_β of the end of the load application (except at $S_r = 1$); the retardation of maximum pore pressures is the longer, the smaller the saturation degree.
(b) At the same layer thickness and the same saturation degree, the excess pore-pressure versus the total pressure interval ratio is the smaller, the smaller the initial porosity and the smaller the load interval.
(c) At equal load interval and equal final total pressure the consolidation lines $e = e(t)$ tend towards the same asymptotic line of the secondary consolidation irrespective of the initial void ratio and saturation degree. For the chosen isotache set, this line is a logarithmic straight line according to the equation

$$e = e_o - \alpha_e \log(t/t_o)$$

(d) At the same value t/t_β, the void ratio decrease is the more important, the smaller the saturation degree and the thicker the layer. Furthermore, if e_α denotes the initial void ratio and $e_{\omega q}$ the void ratio in a very great time t_ω after the beginning of the application of the load q, then at a time $t_\beta < t < t_\omega$ the quotient $(e_\alpha - e) : (e - e_{\omega q})$ is the smaller, the smaller the load interval q.

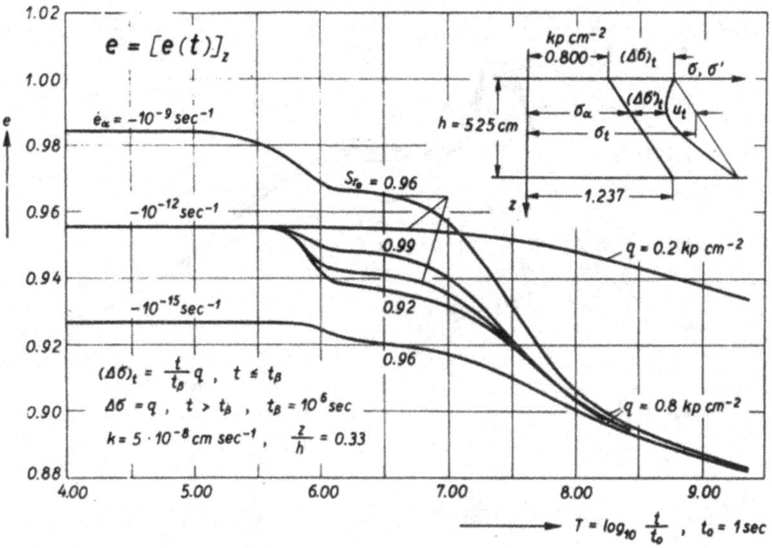

Fig. 10.2 - Comparative consolidation curves for variations in S_{ro}, e_{α} and q

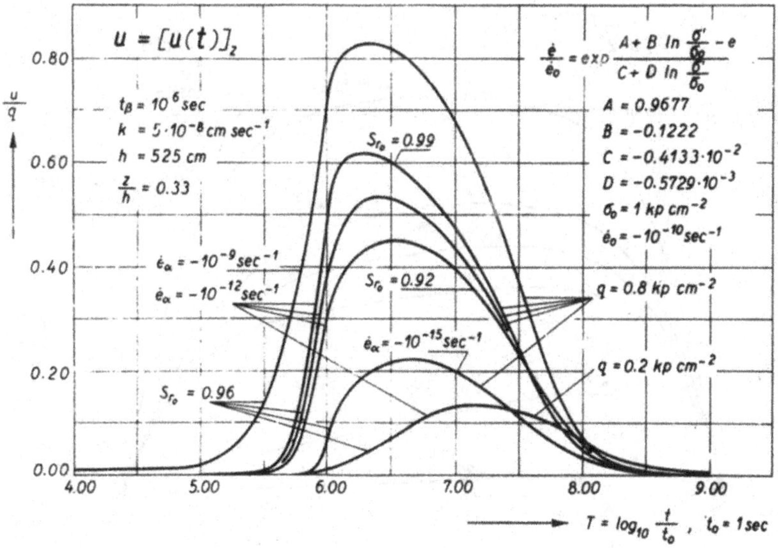

Fig. 10.3 - Comparative pore-pressure versus time plots for variations in S_{ro}, $\dot{e}_{\alpha}$ and q

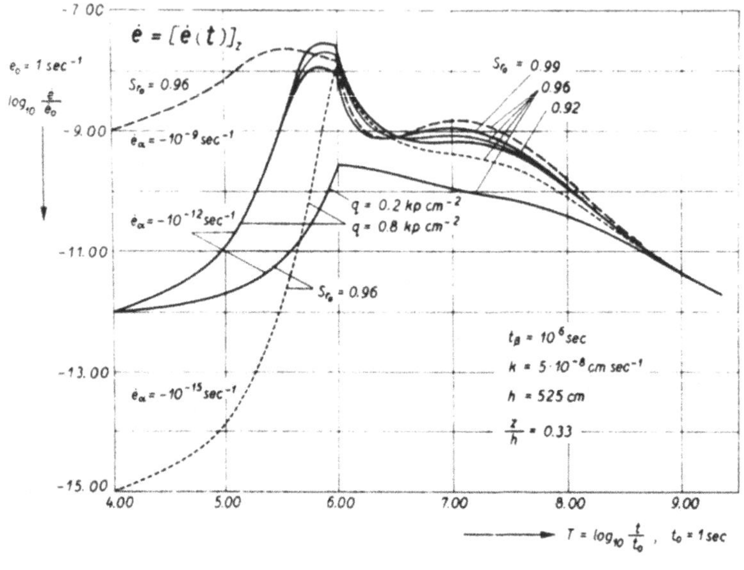

Fig. 10.4 - Comparative void ratio speed versus time plots for variations in S_{ro}, $\dot{e}_\alpha$ and q

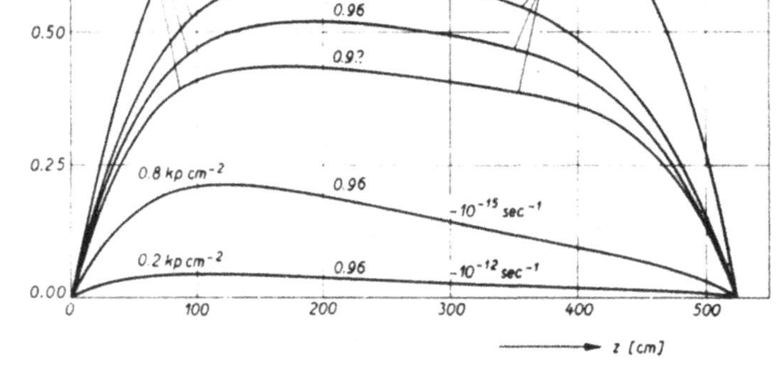

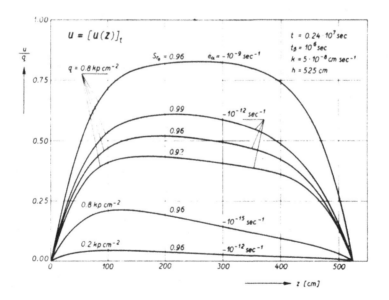

Fig. 10.5 - Comparative isochrones for variations in S_{ro}, e_α and q

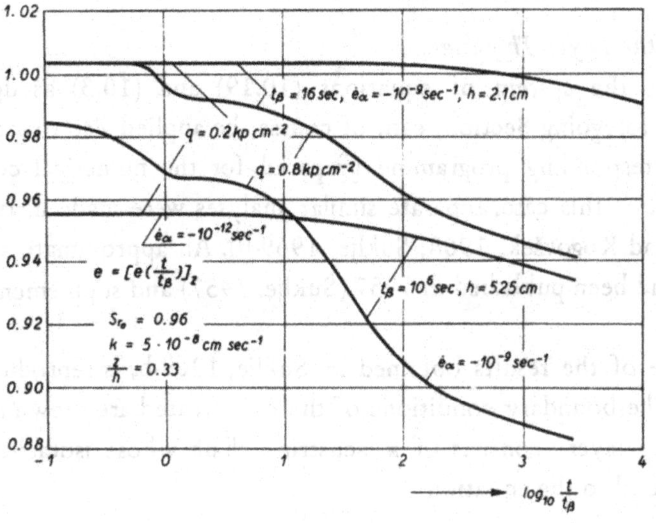

Fig. 10.6 - Comparative consolidation curves for variations in h and q

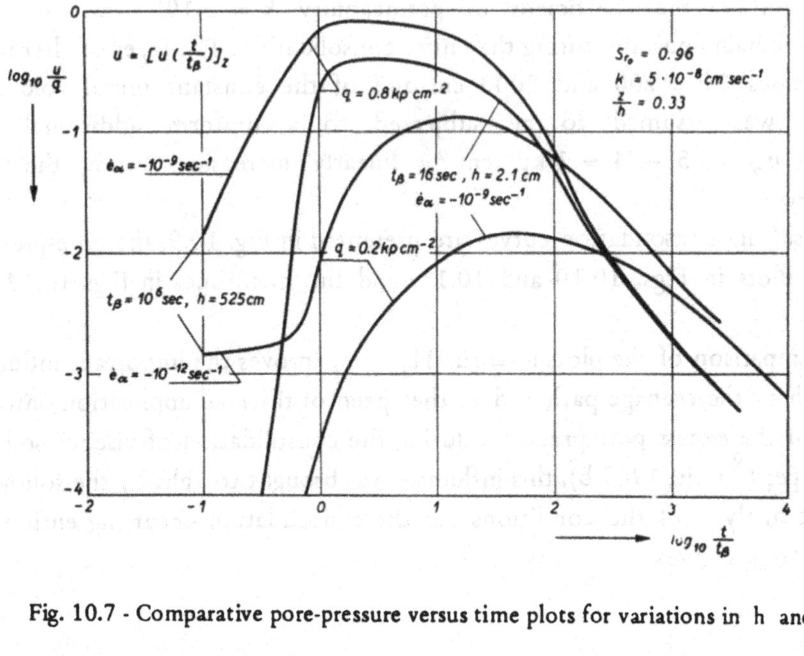

Fig. 10.7 - Comparative pore-pressure versus time plots for variations in h and q

One-Dimensional Consolidation of Saturated Soils

(a) Influence of the Layer Thickness

With $b = 0$, the system of equations (10.19) and (10.3) as deduced and presented in the foregoing Section, can, of course, be applied also to saturated soils ($S_r = 1$); the corresponding programme prepared for the numerical computation, remains in use. For this case, separate similar analyses were made in two previous papers (Šuklje and Kogovšek, 1968; Šuklje, 1969-b). An approximate study of the same problem had been published in 1957 (Šuklje, 1957) and supplemented in 1969 (Šuklje, 1969-a).

An example of the results obtained by Šuklje, 1969-b, is reproduced in Figs. 10.8 to 10.13. The boundary conditions of the case treated are shown in Fig. 10.8. The consolidating layer consists of a lacustrine chalk whose isotaches had been found to correspond to the equation

$$(10.32) \qquad -\dot{e} = \exp\left[2.3026\left\{(A + B\,e) + \left[(C + D\,e)\frac{\sigma'}{\sigma_o}\right]^{-1}\right\}\right]$$

with the parameters $A = -81.11$, $B = 137.50$, $C = 0.03820$, $D = -0.09125$, $\sigma_o = 1\ \text{kp cm}^{-2}$; the coefficient of permeability $k = 3.10^{-7}\ \text{cm sec}^{-1}$ was supposed to remain constant during the entire consolidation. The layer of alternative thickness values $2h = 200$ and 2000 cm and of the constant initial void ratio $e_\alpha = 0.615$, was assumed to be subjected to a uniform additional load $\Delta\sigma = \sigma_\beta - \sigma_\alpha = 5 - 4 = 1\ \text{kp cm}^{-2}$, linearly increasing during the time $t_\beta = 10^7 \text{sec}$.

The resulting consolidation curves are presented in Fig. 10.9, the pore-pressure versus time plots in Figs. 10.10 and 10.11, and the isochrones in Figs 10.12 and 10.13.

The comparison of the plots $u = [u(t)]_{z = \text{const}}$ proves the important influence of the length of the seepage path and of the speed of the load application onto the magnitude of the excess pore-pressures during the consolidation of viscous soils. In the same paper (Šuklje, 1969-b), this influence was brought to light by the following approximate analysis of the conditions for the consolidation occurring entirely at very small pore-pressures.

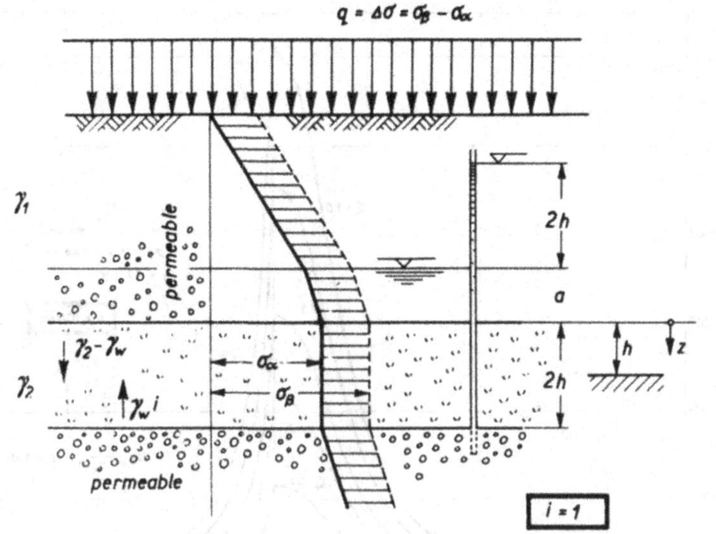

Fig. 10.8 - Boundary conditions for consolidation lines, pore-pressure versus time plots and isochrones presented in Figs. 10.9 to 10.13

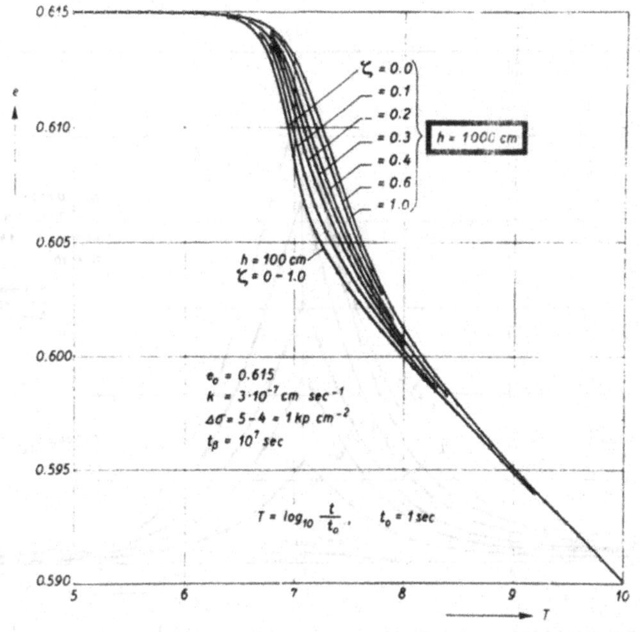

Fig. 10.9 - Consolidation lines corresponding to the boundary conditions of Fig. 10.8

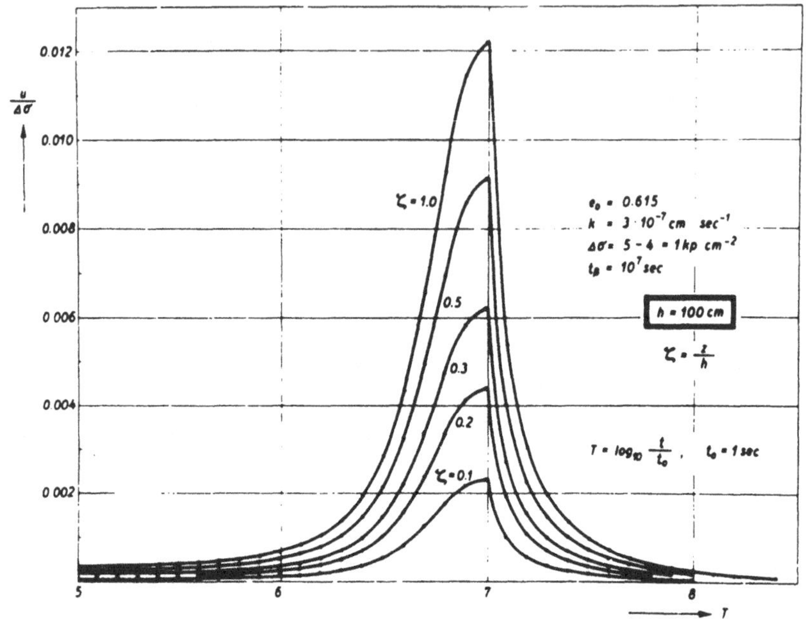

Fig. 10.10 - Pore-pressure versus time plots corresponding to the boundary conditions of Fig. 10.8, h = 100 cm

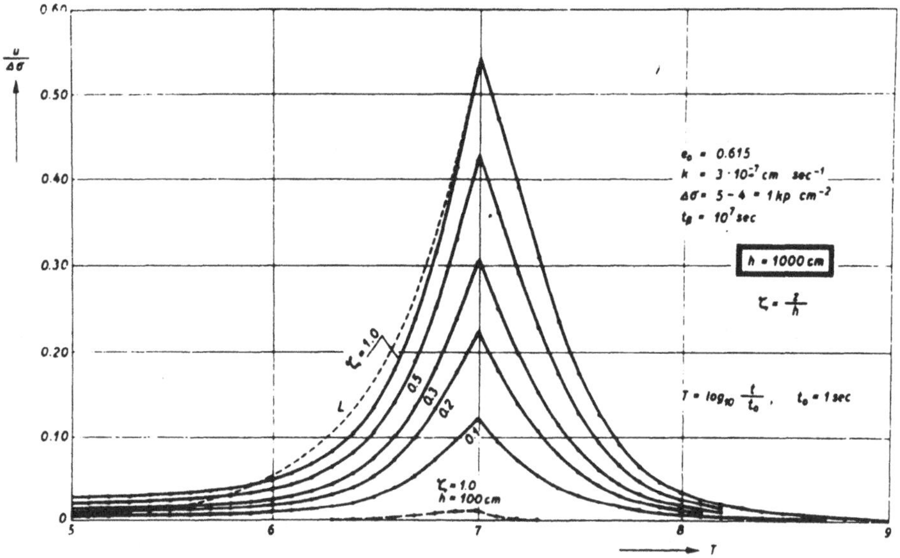

Fig. 10.11 - Pore-pressure versus time plots corresponding to the boundary conditions of Fig. 10.8, h = 1000 cm

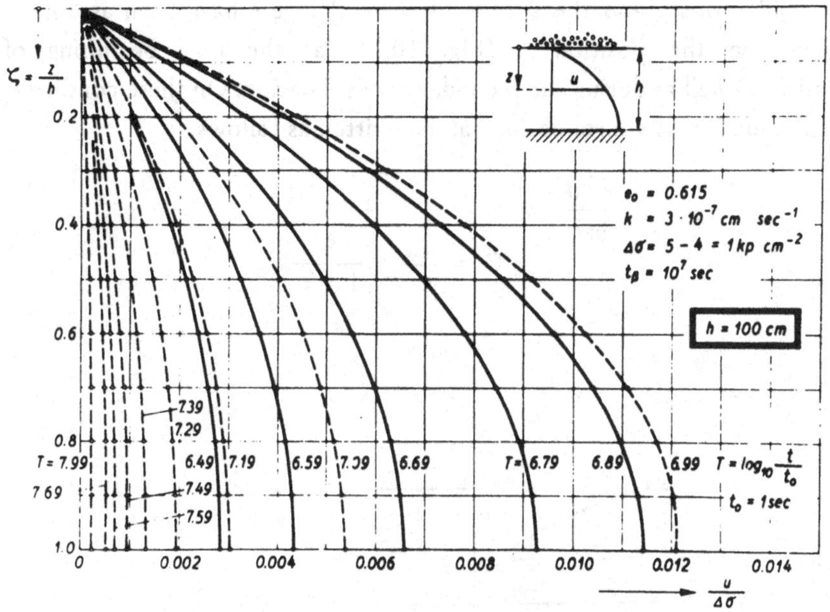

Fig. 10.12 - Isochrones corresponding to the pore-pressure versus time plots presented in Fig. 10.10

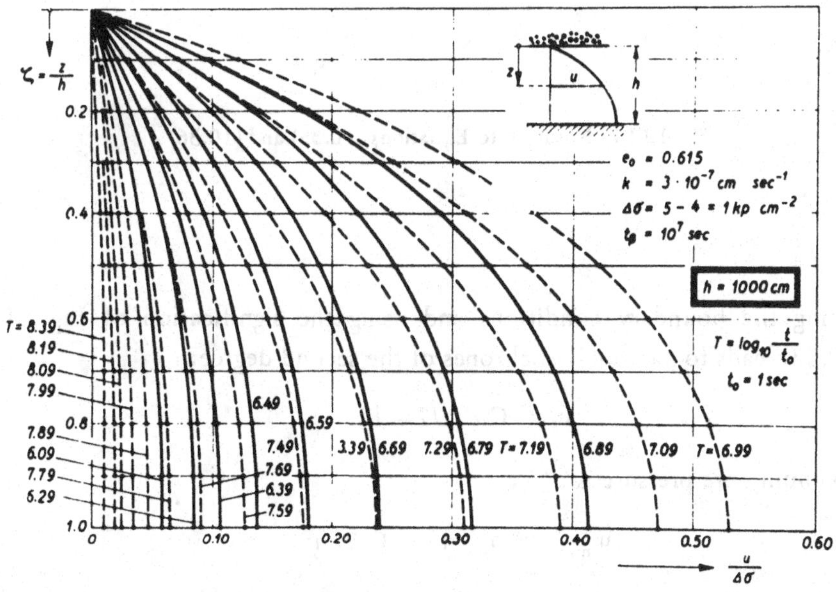

Fig. 10.13 - Isochrones corresponding to the pore-pressure versus time plots presented in Fig. 10.11

The void ratios e and the speeds of their change $\dot{e}$ have been assumed to be independent on the position z (Fig. 10.14) at the very beginning of the consolidation as well as during the secondary consolidation. On these conditions, the differential equation of consolidation can be written as follows:

(10.33) $$\frac{\partial^2 u}{\partial z^2} = C$$

(10.34) $$C = \frac{\gamma_w}{k} \cdot \frac{\dot{e}}{1 + e}$$

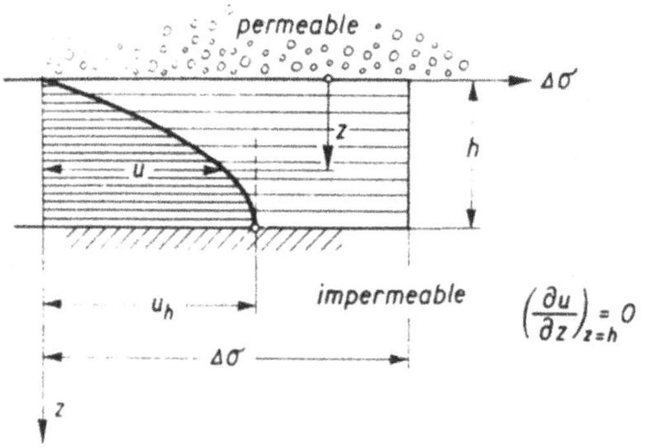

Fig. 10.14 - Notation to Equations (10.35) and (10.36)

Considering the boundary conditions and using the signification of Fig. 10.14, Equ. (10.33) leads to parabolic isochrones of the second degree:

(10.35) $$u = C\, z\, (z/2 - h)$$

The maximum pore pressure is

(10.36) $$u_{max} = u_h = -(C/2)\, h^2$$

Consequently, the pore pressures will not exceed a certain small value u_s provided the length of the filtration path is smaller than h_o:

$$h \leqslant h_o = \sqrt{\left(\frac{2u_h}{C}\right)} = \sqrt{\left(-\frac{2k(1+e)}{\gamma_w \dot{e}} u_s\right)} \tag{10.37}$$

Using the equation of isotaches

$$\dot{e} = \dot{e}(e, \sigma')$$

we can obtain, from Equation (8.37), the line

$$e = [\, e(\sigma')\,]_{h_o, u_s} \tag{10.38}$$

(Fig. 10.15). If the consolidation path $e = e(\sigma')$ lies below this line, the consolidation of all the layers whose thickness is smaller than or equal to h_o, occurs at pore pressures that are smaller than u_s. If the starting point (e_o, σ'_o) of the consolidation path (Fig. 10.15) lies below the line (10.38), the consolidation path of

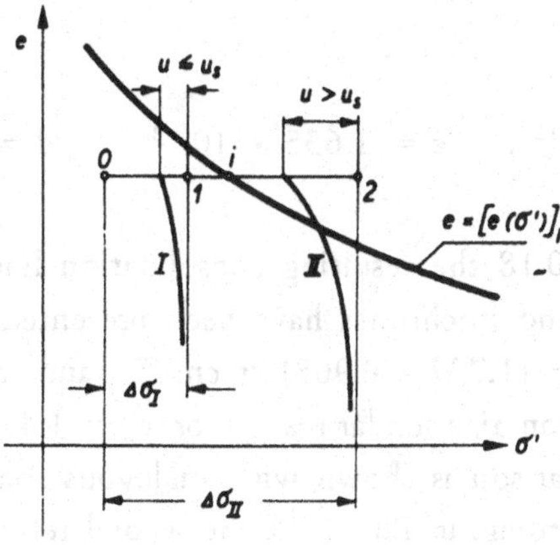

Fig. 10.15 - Consolidation paths I and II and the plot $e = [e(\sigma')]_{h_o, u_s}$

a layer $h \leqslant h_o$ (I in Fig. 10.15) will certainly occur at pore pressures less than u_s

provided that the total stress increase $\Delta\sigma_I$ = 0–1 is less than the distance 0–i of the point 0 from the line (10.38) on the level e_o = const. If, however, the total stress increase is greater ($\Delta\sigma_{II} = 0-2 > 0-i$ in Fig. 10.15), the position of the initial part of the consolidation path above (II in Fig. 10.15) or below the line (10.38) will depend on the magnitude of the part i–2 of the load increase, on the speed of the load increase and on the coefficient of permeability. The condition $u \leqslant u_s$ will also be assured in this case provided the speed of the total stress increase does not surpasse a certain limit value. (Considerations concerning this limit value have been presented in the paper by Šuklje, 1969-b).

(b) Role of the Effective Stress Speed

For the rheological relationship containing the factor $\dot{\sigma}'$ (speed of the effective stress change) in the form of Eq. (2.7)

$$- \dot{e} = c\,e\,\{ t_0^{-1} (e/e_0)^{1/c} (\sigma'/\sigma_0)^{1/d} + [\dot{\sigma}'/(\sigma' d)]\,[1 - (e/e_0)^{1/c}]\}$$

($t_0^{-1} = A\,e_0^{\;1/c}\,\sigma_0^{1/d}$), Šuklje and Kovačič (1974) presented examples of the numerical solution of the differential equation of the one-dimensional consolidation. The boundary conditions were taken according to Fig. 10.8 and (using the time lines presented in Fig. 4 of Garlanger's paper, 1972) the following rheological parameters considered (kp, cm, sec):

$$A = 3.83 \times 10^{-15}, \qquad c = 1.635 \times 10^{-2}, \qquad d = 5.484 \times 10^{-2}.$$

In Figs. 10.16 to 10.18 the resulting consolidation lines, pore-pressure ratio versus time ratio plots, and isochrones have been presented (lines b) for the load interval $\Delta\sigma = \sigma_\beta - \sigma_\alpha = (1.737 - 0.908)$ kp cm^{-2} and the initial void ratio $e_0 = 1.294$. (The paper contains similar results for $e_0 = 1.4139$.)

In Fig. 10.17 comparison is shown with analogous solutions (lines a) which were obtained by disregarding, in Eq. (2.7), the second term containing the factor $\dot{\sigma}'$. There are small differences here. The examination of numerical results proves, however, that they have to be attributed to the high sensitivity of the quotient e/e_0 when raising it to the power $1/c$. As to the consolidation lines (Fig. 10.16), the differences are too small to be expressed graphically. The isochrones in Fig. 10.18 correspond to the lines (a) in Fig. 10.17.

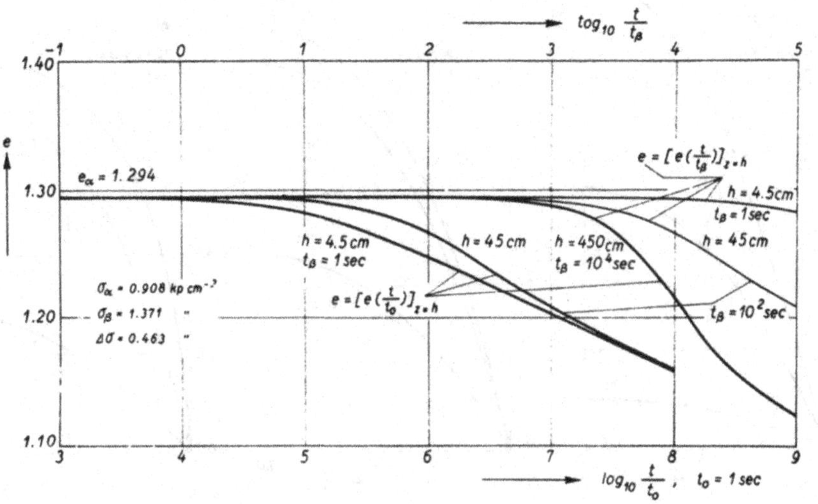

Fig. 10.16 - Consolidation lines

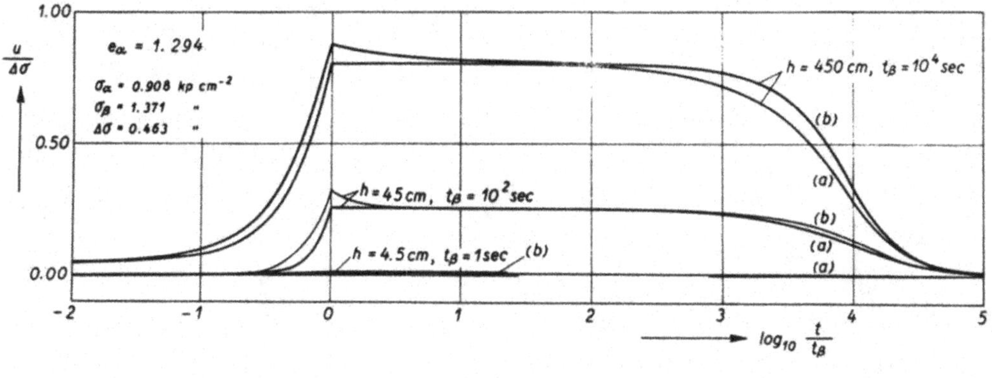

Fig. 10.17 - Pore-pressure versus time plots: (a) disregarding the second term in Eq. (2.7), (b) considering both terms

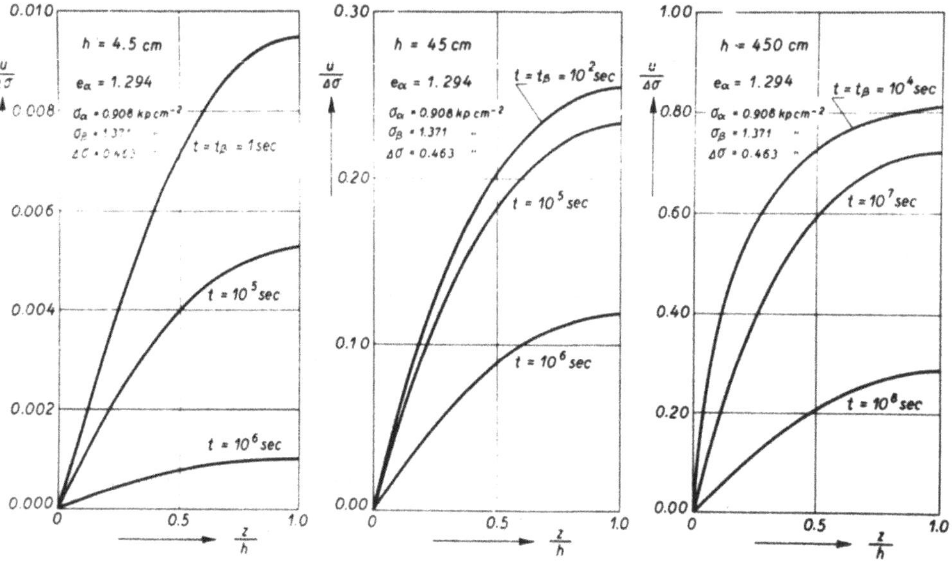

Fig. 10.18 - Isochrones. Figs. 10.16 to 10.18 reproduced after Šuklje and Kovačič, 1974. Boundary conditions according to Fig. 10.8

11. TWO-DIMENSIONAL CONSOLIDATION ANALYSIS FOR NON-LINEAR VISCOUS SOILS

Combined Radial and Vertical Consolidation of Saturated Multi-Layer Soils

Until the stress state under an embankment approaches failure, the vertical normal stress distribution does not appreciably change during the consolidation process. When, furthermore, the thickness of consolidating layers is much smaller than the width of the embankment, the horizontal displacements accompanying the compaction (i.e. following the initial "undrained" displacements) under the central part of the load surface are not important. In such cases the rheological relationships corresponding to oedometer test data can be used in the consolidation analysis with acceptable accuracy. The vertical total pressures can be considered as known values at any time during consolidation and the consolidation analysis reduces to the solution of the diffusion equation for given boundary conditions.

Šuklje and Kovačič (1978) have elaborated a numerical solution of the diffusion equation for multi-layer viscous soils that are subjected to gradually increasing embankment loads and the consolidation of which is accelerated by a system of vertical drains. The program for the numerical computation has been applied to a test embankment and the results added to "undrained" settlements (see Chapter 4: Examples, b) have been compared with field observations.

The diffusion equation has been applied in the form (see Eq. 9.42) adapted to saturated soils with zero limit hydraulic gradients, with hydrostatic initial pore-water potential field (neglecting the very small excess pore-pressures governing the previous secondary consolidation) and to constant ratio κ of the permeability anisotropy:

$$\kappa = k_r/k_z = k_r/k = \text{const} \tag{11.1}$$

The coefficient of permeability $k = k_z$ has been considered as a function of the void ratio:

$$k = k(e) \tag{11.2}$$

The results of permeability tests performed in oedometer may be presented by a set of point values and splines (Desai, 1971) used for interpolation. Alternatively, the relationship $k = k(e)$ can be given an appropriate analytical expression (see e.g. Eqs. 2.18). The coefficients $\kappa = k_r/k$ can be obtained by testing also samples whose axis corresponds to the horizontal direction in situ.

The adaptation of the diffusion equation to the above requirements yields:

$$(11.3) \qquad \frac{\partial e}{\partial t} = \frac{1+e}{\gamma_w} \{ \kappa [k \frac{\partial^2 u}{\partial r^2} + (\frac{\partial k}{\partial r} + \frac{k}{r}) \frac{\partial u}{\partial r}] + k \frac{\partial^2 u}{\partial z^2} + \frac{\partial k}{\partial z} \frac{\partial u}{\partial z} \}$$

(cf. Eq. 9.58). Rheological relationships have been taken into account in the form corresponding to the non-linear Kelvin body according to Eq. (1.20):

$$(1.20) \qquad e = e_0(\sigma') - A(\sigma') \{ 1 - (\dot{e}/\dot{e}_0)^n \}$$

The stress dependent parameters $e_0(\sigma')$ and $A(\sigma')$ have been presented by sets of point values and splines used for interpolation. The principle of effective stress has been considered in the form

$$(11.4) \qquad \sigma' = \sigma_r - u$$

$$(11.5) \qquad u = u^* - u_{in}$$

u^* = total pore-pressure, u_{in} = pore-pressure corresponding to the initial hydrostatic pore-pressure field (if disregarding the small excess pore-pressures governing the previous secondary consolidation), σ_r = the reduced value of the total vertical pressure:

$$(11.6) \qquad \sigma_r = \sigma - u_{in}$$

σ = total vertical pressure.

A computer program has been prepared enabling the numerical integration of the diffusion equation (11.3) for boundary conditions presented in Fig. 11.1. The drains as well as the efficiency domain of any single drain are assumed to be cylindrically shaped with the inner radius r_0 and outer radius R. The consolidation

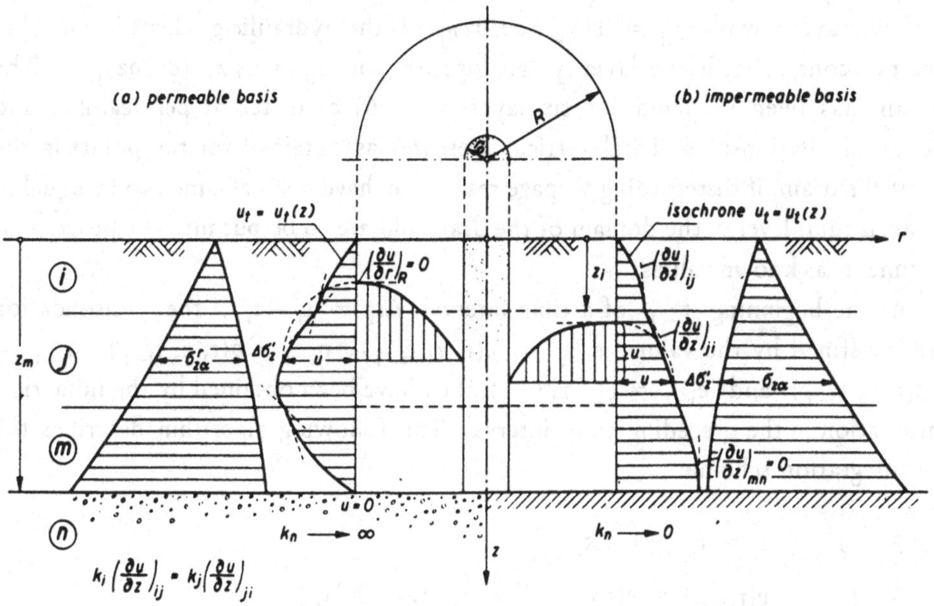

Fig. 11.1 - Boundary conditions for the consolidation analysis

of the soil below the bottom level of drains can be considered by taking $\kappa \to 0$. According to Fig. 11.1 the following boundary conditions for excess pore-pressures have been taken into account.

$$0 \leqslant z \leqslant z_m \quad \begin{cases} r = r_0 & u = 0 \\ r = R & \dfrac{\partial u}{\partial r} = 0 \end{cases} \qquad \begin{matrix} (11.7) \\ (11.8) \end{matrix}$$

$$r_0 \leqslant r \leqslant R \quad \begin{cases} z = 0 & u = 0 \\ z = z_m & \begin{cases} \text{(a)} & \dfrac{\partial u}{\partial z} = 0 \\ \text{(b)} & u = 0 \end{cases} \end{cases} \qquad \begin{matrix} (11.9) \\ (11.10\text{-a}) \\ (11.10\text{-b}) \end{matrix}$$

Furthermore, the interlayer conditions of the continuity of water flow have to be considered:

$$k_i \left(\frac{\partial u}{\partial z}\right)_{ij} = k_j \left(\frac{\partial u}{\partial z}\right)_{ji} \qquad (11.11)$$

k_i, k_j are permeability coefficients $k = k_z$ of layers i and j resp. at the contact of these two layers, while $i_{ij} = 1/\gamma_w\ (\partial u/\partial z)_{ij}$ is the hydraulic gradient in the layer i at the contact with the layer j (analogously for $i_{ji} = 1/\gamma_w\ (\partial u/\partial z)_{ji}$). The program has been prepared for six layers ($m = 6$) of different permeability and rheological relationships. Total vertical pressures as obtained for the points in the axis of the drain, if disregarding seepage resistance, have been assumed to be equal in any horizontal level in the domain of the drain and are to be put into the program at any time t as known values.

At the beginning t_{i-1} of a time interval $\Delta t_i = t_i - t_{i-1}$ the consolidation state is defined by the values $e_{i-1} = e(r, z, t_{i-1})$, $\sigma_{i-1} = \sigma(r, z, t_{i-1})$, $u_{i-1} = u(r, z, t_{i-1})$ and $\dot{e}_{i-1} = \dot{e}(r, z, t_{i-1})$; they have been obtained by the numerical computation in the preceding time interval. The following algorithm describes the time integration scheme:

(11.12) (1) $t_i = t_{i-1} + \Delta t_i$

(11.13) (2) $e(r,z,t_i) = e(r,z,t_{i-1}) + \dot{e}(r,z,t_{i-1})\,\Delta t_i$

(3) Compute $\sigma(r,z,t_i)$.

(4) Solve the boundary value problem defined by the combination of relations (11.3) and (1.20), boundary conditions (11.7), (11.8), (11.9), (11.10-a) or (11.10-b), and interlayer conditions (11.11). Differential equation

$$\kappa[k\frac{\partial^2 u}{\partial r^2} + (\frac{\partial k}{\partial r} + \frac{k}{r})\frac{\partial u}{\partial r}] + k\frac{\partial^2 u}{\partial z^2} + \frac{\partial k}{\partial z}\frac{\partial u}{\partial z} = \frac{\gamma_w\,\dot{e}_0}{1+e}\,[1 + \frac{e - e_0(\sigma')}{A(\sigma')}]^{1/n}$$

(11.14)

is non-linear because u takes part in relations $e_0(\sigma')$ and $A(\sigma')$ on the righthand side.

(5) Compute $e(r, z, t_i)$ from Eq. (1.20).

(6) If $t < t_{end}$, continue step (1), otherwise procedure is completed.

The boundary value problem in step (4) can be solved by the finite difference method. Rectangular region in the (r, z) space defined by the inner radius r_0, outer radius R and layer depths $h_1, \dots h_n$ is subdivided in r and z directions and forms a rectangular mesh. By substituting all partial derivatives by differential quotients, and considering all boundary and inter-layer conditions, we get a system of non-linear equations of the type

$$\sum_i a_{ik}\, u_k = f(u_k)\,, \qquad i = 1,2,\ldots n = \text{number of equations} \qquad (11.15)$$

Newton's method is recommended because good initial approximation for u can be obtained by linear extrapolation of two former values in time.

(For some further data concerning the realization in computer program see Šuklje and Kovačič, 1978.)

Some results of the application of the program for forecasting settlements of a test embankment (64m × 190m) constructed on very compressible lacustrine sediments, are presented in Figs. 11.3 and 11.4 (reproduced after Šuklje and Kovačič, 1978). The cross-section through the embankment has been shown in Fig. 8.4, and the sequence of five silty and clayey layers underlying the surface peat and overlying the permeable rocky base is denoted in Fig. 11.3.

The consolidation of the upper three layers of 11.25 m thickness has been accelerated by sandy drains (with too large mutual distances of 2.30 m to be sufficiently effective). As an example of oedometer test data used in the consolidation analysis, Fig. 11.2 presents the plots $e_0 = e_0(\sigma')$, $\alpha = \alpha(\sigma')$ and

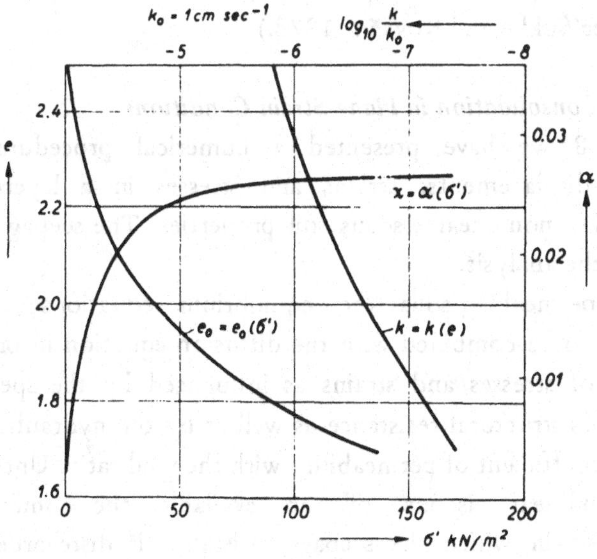

Fig. 11.2 - Isotache $e_0 = e_0(\sigma')$ for $\dot{e} = \dot{e}_0 = -10^{-7}\ \text{sec}^{-1}$, plot $\alpha = \alpha(\sigma')$ and plot $k = k(e)$ for the silty layer MH

$k = k(e)$ corresponding to the upper silty layer MH (Fig. 11.3).

Coefficients α and n determine the parameter $A(\sigma')$ according to Eq. (1.30). The consolidation analysis has been made for the following values of parameters n and κ and of the speed of the previous secondary consolidation:

$n = 0.002$ for layers MH, CH/CI and CI/MI, $n = 0.2$ for layers CH/MH and CI/MI ;
$\kappa = 3$ for layer MH, $\kappa = 1$ for other layers;
$\dot{e}_\alpha = -10^{-14}\ \text{sec}^{-1}$ for layer MH, $\dot{e}_\alpha = -10^{-12}\ \text{sec}^{-1}$ for other layers.

In Fig. 11.3 the resulting isochrones of excess pore-pressures are compared with some measured values as well as with the results which correspond to the following assumptions: $n = 0$ (Buisman's law of the secondary consolidation), $\kappa = 1$ for all layers, $\dot{e}_\alpha = -10^{-10}\text{sec}^{-1}$ for all layers.

The resulting consolidation line (line 1 in Fig. 11.4) has been combined with the line of undrained settlements (line 2) as obtained by the numerical computation explained in Chapter 2 (see also section "Examples (b)" in Chapter 2). The combined settlement line 3 has been compared, in Fig. 11.4, with the line 4 of measured settlements. (For the interpretation of differences between computed and measured values see Šuklje and Kovačič, 1978.)

Two-Dimensional Consolidation in Plane-Strain Conditions

In Chapter 8 we have presented a numerical procedure enabling the computation of displacements, strains and stresses in a layered plane-strain half-space exhibiting non-linear viscous soil properties. The seepage resistance was not considered in the analysis.

In poorly permeable soils the equilibrium equations and rheological relationships have to be combined with the diffusion equation in order to forecast the development of stresses and strains as influenced by the speed of the load increase and viscous structural resistance as well as by the hydraulic resistance and the change of the coefficient of permeability with the void ratio. Until such complex solutions for non-linear viscous soils are available, the combination of the stress-strain analysis in which the seepage resistance is disregarded, and of the separate solution of the diffusion equation can serve for an approximate prediction of the development of settlements. Thereby the total stresses as obtained by the "drained" analysis, have to be considered known values at any time during the consolidation.

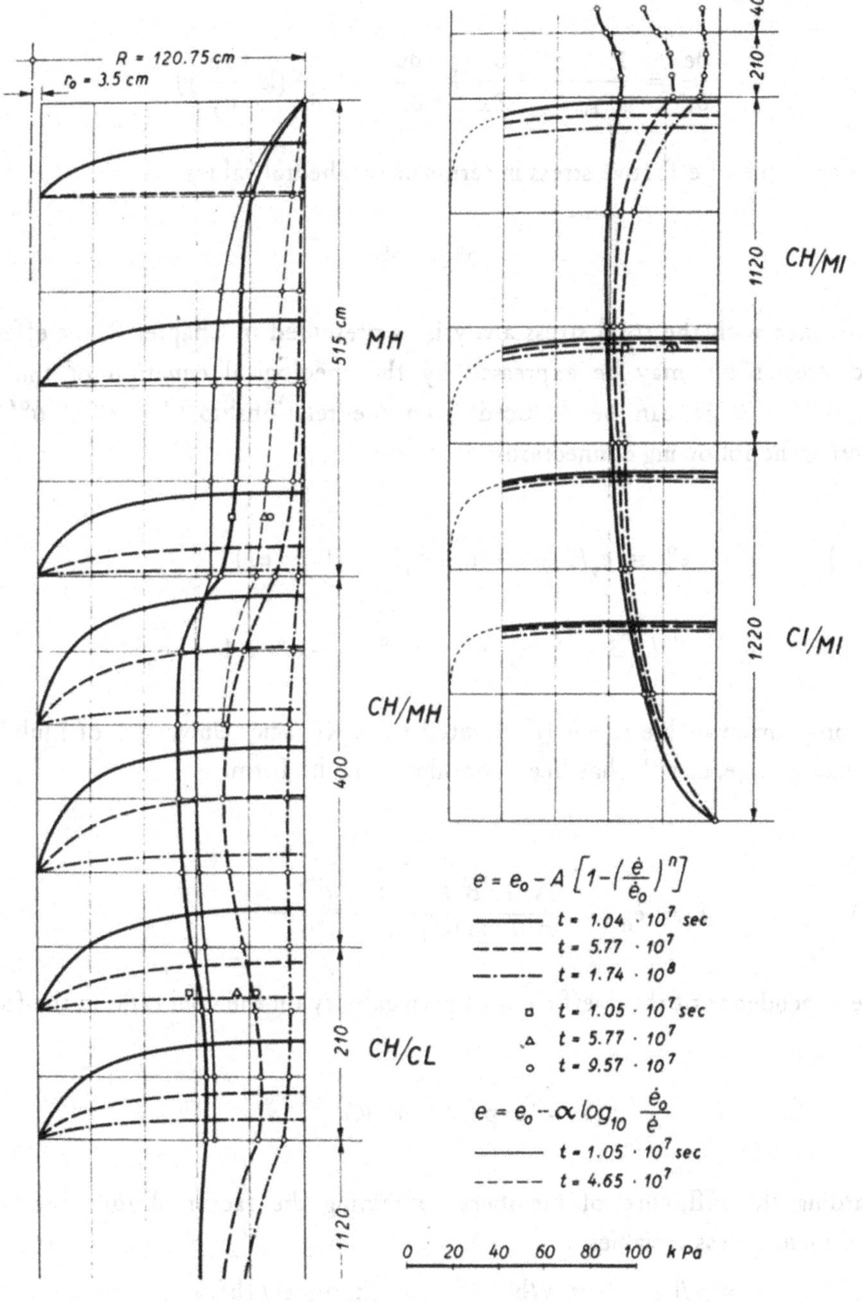

Fig. 11.3 - Isochrones in the axis of the embankment (see Fig. 8.4) compared to some measured pore pressures

The diffusion equation may be applied in the form (see Eq. (9.34))

$$\frac{\partial e}{\partial t} = \frac{1 + e}{\gamma_w} \left\{ \frac{\partial}{\partial x}\left(k_x \frac{\partial u}{\partial x}\right) + \frac{\partial}{\partial y}\left(k_y \frac{\partial u}{\partial y}\right)\right\} \tag{11.16}$$

and the principle of effective stress in terms of octahedral values

$$u = \sigma^o - \sigma^{o\prime} \tag{11.17}$$

In accordance with the total stress analysis as presented in Chapter 8, the effective spheric stresses $\sigma^{o\prime}$ may be expressed by the rheological equation of the type $R(e, \dot{e}, \sigma^{o\prime}) = 0$. It can be deduced from the relationship $\epsilon^o = \epsilon^o(\epsilon^o, \sigma^{o\prime})$ by considering the following connections:

$$\epsilon^o = \epsilon_v/3 = -(e - e_0)/\{3(1 + e_0)\} \tag{11.18-a}$$

$$\beta_o = \beta_e/\{3(1 + e_0)\} \qquad \dot{\epsilon}^o = -\dot{e}/\{3(1 + e_0)\} \tag{11.18-b,c}$$

In the program available so far (elaborated by I. Kovačič, University of Ljubljana) the equation $R(e, \dot{e}, \sigma^{o\prime})$ has been considered in the form.

$$\dot{e} = \dot{e}_0 \exp \frac{A + B \ln(c + d\,\sigma^{o\prime}) - e}{C + D \ln(c + d\,\sigma^{o\prime})} \tag{11.19}$$

and the dependence of the coefficient of permeability on the void ratio in the form

$$k = \exp(M e + R)$$

Disregarding the influence of members containing the factor $\partial k/\partial e$ and introducing dimensionless variables

$$\tilde{x} = x/l\,, \quad \tilde{y} = y/h\,, \quad \tilde{t} = (k_0\, u_0\, t)/(h^2\, \gamma_w)$$
$$\tilde{u} = u/u_0\,, \quad \tilde{k} = k/k_0$$

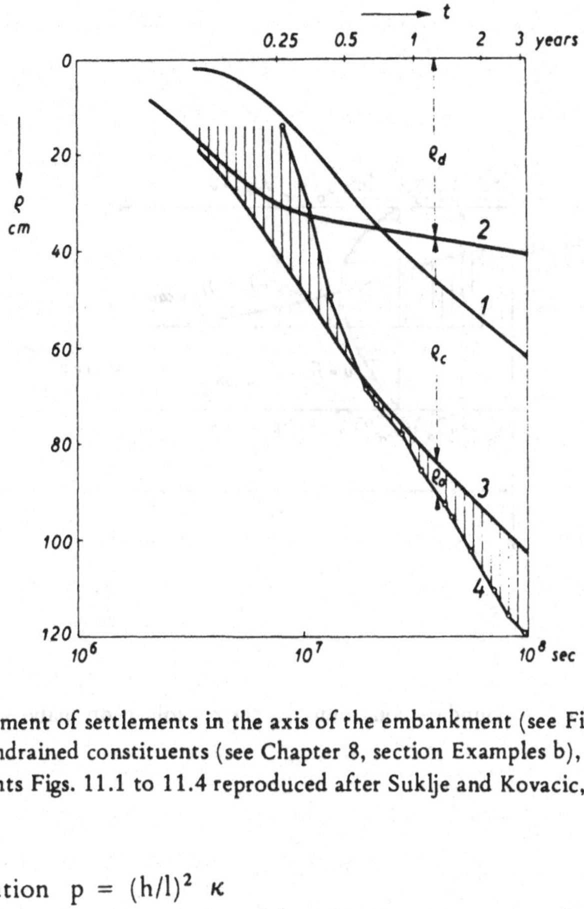

Fig. 11.4 - Development of settlements in the axis of the embankment (see Fig. 8.4): (1) consolidation constituents, (2) undrained constituents (see Chapter 8, section Examples b), (3) = (1) + (2), (4) measured settlements Figs. 11.1 to 11.4 reproduced after Suklje and Kovacic, 1978.

and the substitution $p = (h/l)^2 \ \kappa$

the problem is expressed in the dimensionless form:

$$p \frac{\partial^2 \tilde{u}}{\partial \tilde{x}^2} + \frac{\partial^2 \tilde{u}}{\partial \tilde{y}^2} = \frac{\partial e}{\partial \tilde{t}} \ \frac{1}{(1 + e) \tilde{k}(e)} = f(\tilde{u}, \tilde{x}, \tilde{y}) \qquad (11.21)$$

For notation h, l, see Fig. 11.5 presenting the boundary conditions considered in the consolidation analysis (l denotes the distance where the additional total spheric stresses become negligibly small). By applying the method of finite differences the problem was transformed into a system of non-linear equations, which was solved by the Newton method (Šuklje, 1977).

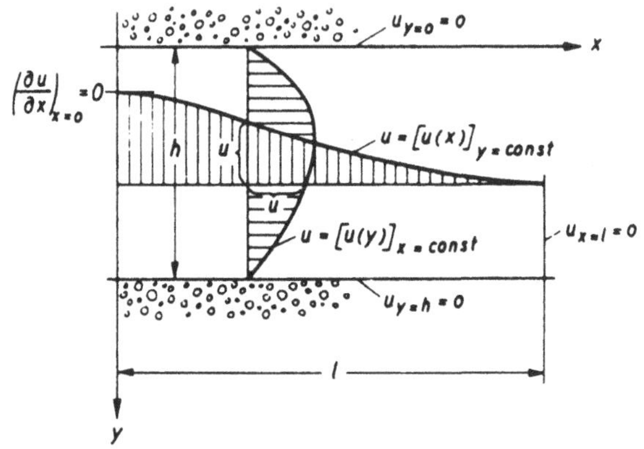

Fig. 11.5 - Boundary conditions for the consolidation analysis

Figs. 11.6 to 11.9 present some results of the application of the above explained computing program to the boundary conditions according to Fig. 8.3. The triangular load strip with maximum intensity $q = 23.53$ kPa was assumed to increase with the velocity $q = 94.14$ kPa year^{-1}. The volume deformability of the soil has been presented by the plots $\epsilon_0^o = \epsilon_0^o(\sigma^{o\prime})$ corresponding to the speed $\dot{e}_0 = -10^{-8}$ sec^{-1} in Fig. 4.2, and by plots $\beta_0 = \beta_0(\sigma^{o\prime})$ in Fig. 4.3. Fig. 11.6 shows some selected excess pore-pressure isochrones, and Fig. 11.7 the pore-pressure versus time plots for two selected points in the layer.

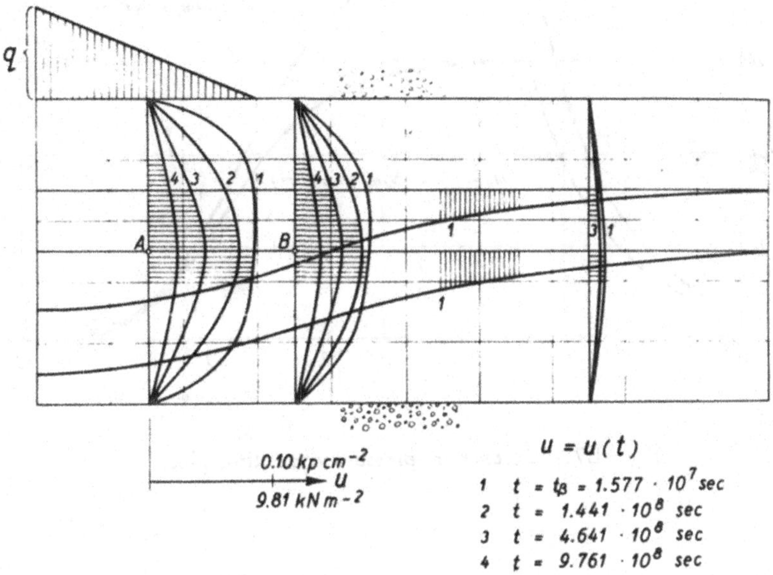

Fig. 11.6 - Isochrones of excess pore-pressures

The consolidation degrees obtained by the numerical solution of the diffusion equation have been considered to govern also the development of differences between total settlements ($v_{K,G}$ in Fig. 8.2) and the pure deviatoric "undrained" settlements (v_{G_0} in Fig. 8.2). The resulting settlement isochrones are shown in Fig. 11.8, and the settlement versus time plots for two selected points in the base of the load strip in Fig. 11.9.

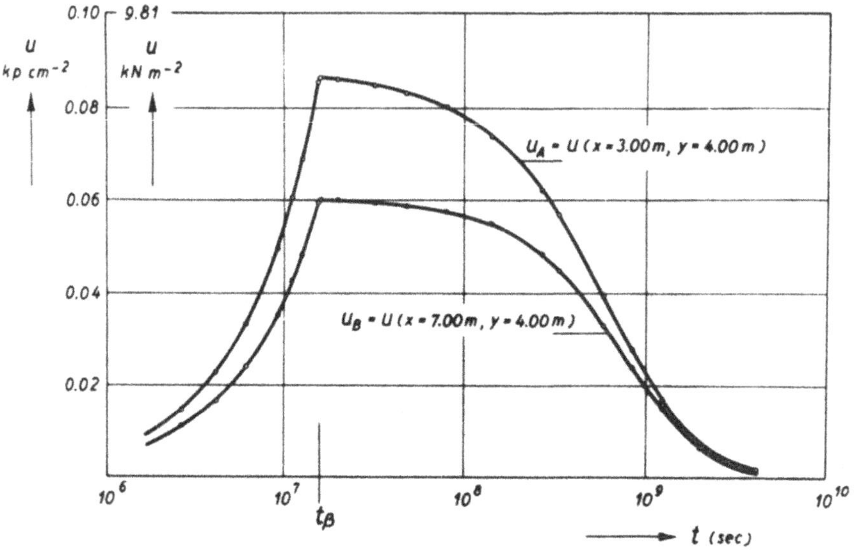

Fig. 11.7 - Excess pore-pressure versus time plots

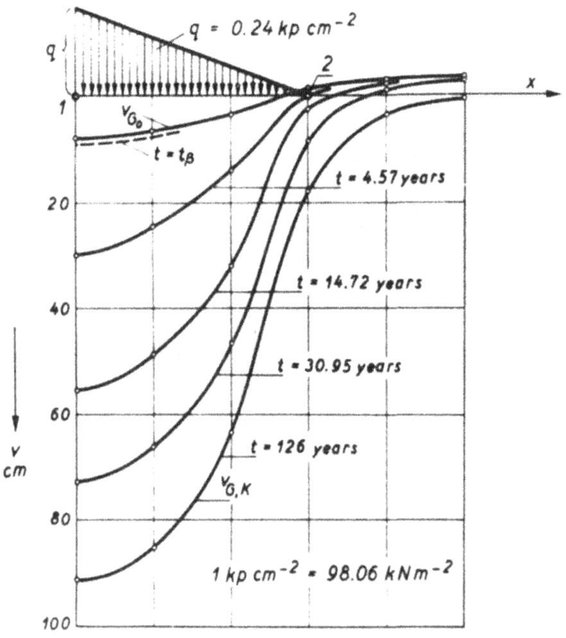

Fig. 11.8 - Settlement isochrones

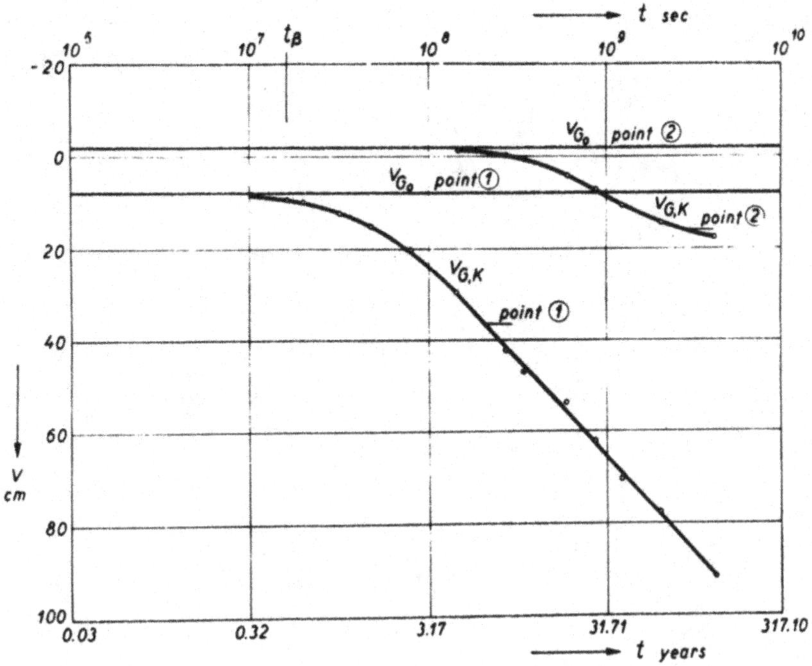

Fig. 11.9 - Settlement versus time plots.
Figs. 11.6 to 11.9 reproduced after Šuklje, 1977

REFERENCES

[1] Anagnosti, P. (1962). Analiza standardnih metoda ispitivanja osobina zemljanih materijala sa stanovista mehanike neprekidnih sredina (The analysis of the standard methods of testing soil properties from the point of view of the mechanics of continua, in Serbo-Croatian). Thesis, University of Beograd.

[2] Barden, L. (1965). Consolidation of clay with non-linear viscosity. Géotechnique 15 : 345-362.

[3] Battelino, D. (1973). Oedometer testing of viscous soils. Proc. 8th Int. Conf. Soi Mech. Found. Eng. Moscow, Vol. 1.1 : 25-30.

[4] Berre T., and Iversen, K. (1972). Oedometer tests with different specimen heights on clay exhibiting large secondary compression. Géotechnique, 22 : 53-70.

[5] Biot, M.A. (1935). Le problème de la consolidation des matières argileuses sous un charge. Anls Soc. Scient. Brux., B55 : 110-113.

[6] Biot, M.A. (1941). General theory of three-dimensional consolidation. J. Appl. Phys 12 : 155-164.

[7] Bishop, A.W. (1955). The principle of effective stress. Teknisk Ukeblad (Oslo), 106 859-863 (1959). (Also published in Norwegian Geotechnical Institute, Publ. 32 Lecture delivered in Oslo in 1955).

[8] Bjerrum, L. (1967). Seventh Rankine Lecture: Enginering geology of normally consolidated marine clays as related to settlements of buildings. Géotechnique, 17 82-118.

[9] Booker, J.R., and Small, J.C. (1977). Finite element analysis of primary and seconda consolidation. Int. J. Solid Structures, Vol. 13: 137-149.

[10] Buisman, K.A.S. (1936). Results of long duration settlement tests. Proc. Int. Co Soil Mech. Found. Eng., Harvard University, 1 : 103-106.

[11] Chang, T.Y., Ko, H.Y., Scott, R.F., and Westmann, R.A. (1968). Granular Materials, Nonlinear characterization, II: Nonlinear analysis. Caltech, Pasadena.

12] Darve, F. (1974). Contribution à la determination de la loi rhéologique incrémentale des sols. Thèse, L'Université Scientifique et Médicale de Grenoble, 176 p.

13] de Josselin de Jong, G. (1957). Application of stress functions to consolidation problems. 4th Int. Conf. Soil Mech. Found. Eng., London, Vol. I : 320-323.

[14] de Josselin de Jong, G. (1968). Consolidation models consisting of an assembly of viscous elements or a cavity channel network. Géotechnique, 18: 195-228.

[15] Desai, C.S. (1971). Nonlinear analyses using spline functions. J. Soil Mech. Found. Div., ASCE, Vol. 97, SM 10.

[16] Desai, C.S. (1972). State-of-the-art: Overview, trends and projections: Theory and application of the finite element method in geotechnical engineering. US Army Eng. Waterways Exp. Station, Corps of Engineers, Vicksburg, Mississippi, pp. 3-90.

17] Drucker, D.C. (1964). Concept of path independence and material stability for soils. Rheology and Soil Mechanics. IUTAM Symposium Grenoble 1964, 23-46, Springer-V., Berlin.

[18] Duncan, J.M., and Chang, Chin-Yung (1970). Nonlinear analysis of stress and strain in soils. J. Soil Mech. Found. Div., ASCE, S5, 1629-1653.

[19] Florin, V.A. (1959, 1961). Osnovy mekhaniki gruntov (Fundamentals of soil mechanics, in Russian): I(356 p.), II (543 p.), Gosstroyizdat, Leningrad.

[20] Folque, J. (1961). Rheological properties of compacted unsaturated soils. Proc. 5th Int. Conf. Soil Mech. Found. Eng., Paris, I: 13-116.

[21] Freudenthal, A.M., and Spillers, W.R. (1964). On the consolidating viscoelastic layer under quasi-static loading. IUTAM Symposium Grenoble 1964: Rheology and Soil Mechanics, Springer-Verlag, Berlin, 196-202.

[22] Garlanger, J.E. (1972). The consolidation of soils exhibiting creep under constant effective stress, Géotechnique 22 : 71-78.

[23] Gibson, R.E., and Lo, K.Y. (1961). A theory of consolidation for soils exhibiting secondary compression. Norwegian Geotechnical Institute, Publ. No. 41.

[24] Gibson, R.E., Schiffman, R.I., and Pu, S.L. (1970). Plane strain and axially symmetric consolidation of a clay layer on a smooth impervious base. Quart. Jour. Mech. Appl. Maths., 23:4: 505-520.

[25] Hansen, Bent (1969). Cit. after Poorooshasb (1969).

[26] Hawley, J.G., and Borin, D.L. (1973). A unified theory for the consolidation of clays. Proc. 8th Int. Conf. Soil Mech. Foun. Eng. Moscow, 1.3: 107-119.

[27] Hwang, C.T., Morgenstern, N.R., and Murray, D.W. (1971). On solutions of plain strain consolidation problems by finite element methods. Canadian Geot. Jour., 8: 109-118.

[28] Janbu, N. (1963). Soil compressibility as determined by oedometer and triaxial tests. Proc. Eur. Conf. Soil Mech. Found. Eng., Wiesbaden, I: 19-26. Discussion, II: 17-21.

[29] Kisiel, I., & Lysik, B. (1966). Zarys reologii gruntów (Outline of the rheology of soils, in Polish). Arkady, Warszawa, 315 p.

[30] Kondner, R.L. (1963). Hyperbolic stress-strain response: Cohesive soils. J. of the Soil Mech. Found. Div., ASCE, Vol. 89, No. SM 1: 115-143.

[31] Kondner, R.L., and Zelasko, J.S. (1963). A hypebolic stress-strain formulation for sands. Proc. 2nd Pan Am. Conf. Soil Mech., Brazil, I : 289-324.

[32] Kulhawy, F.H., Duncan, J.M., and Seed, H.B. (1969). Finite element analysis of stresses and movements in embankments during construction. Report No. TR 70 - 20, Office of Research Services, University of California, Berkeley.

[33] Lade, P.V. (1972). The stress-strain and strength characteristics of cohesionless soils. Thesis. University of California, Berkeley.

[34] Lade, P.V. (1975). Elasto-plastic stress-strain theory for cohesionless soil with curved yield surfaces. UCLA-Eng-7594, November 1975, Los Angeles, 97 p.

[35] Lade, P.V., and Musante, H.M. (1976). Three-dimensional behavior of normally consolidated cohesive soil. UCLA-Eng-7626, April 1976, Los Angeles, 166 p.

[36] Majes, B. (1974). Discussion. Proc. 4th Danube European Conf. Soil Mech. Found. Eng., Bled, II: 68-70.

[37] Mandel, J. (1957). Consolidation des couches d'argiles. Proc. 4th Int. Conf. Soil Mech, Found. Eng., London, 1: 360-367.

[38] Mandel, J. (1961). Tassements produits par la consolidation d'une couche d'argile de grande épaisseur. Proc. 5th Int. Conf. Soil Mech. Found. Eng., Paris, 1: 733-736.

[39] Masson, R. McMillan (1971). Nonlinear characterization and stress analysis in a granular material. Thesis. University of Colorado, Dept. Civ. & Envir. Eng., 222 p.

[40] Mc-Namee, J., and Gibson, R.E. (1960). Plane strain and axially symmetric problems of the consolidation of a semi-infinite clay stratum. Quarterly Journal of Mechanics and Applied Mathematics, 210-227.

[41] Ozawa Yoshio (1973). Elasto-plastic finite element analysis of soil deformation. Dissertation. University of California, Berkeley, 275 p.

[42] Palmerton, J.B. (1972). Creep analysis of Atchafalaya levee foundation. Proceedings of the Symposium on Applications of the finite element method in geotechnical engineering, Vicksburg, Mississippi, 843-862.

[43] Poorooshasb, H.B. (1969). Advances in consolidation theories for clays. Proc. 7th Int. Conf. Soil Mech. Found. Eng., Mexico, 3: 491-497.

[44] Pregl, O. (1974). Die Grundlagen eines Stoffgesetzes für Böden. Mitteilungen des Insitutes für Geotechnik und Verkehrsbau, Hochschule für Bodenkultur, Wien, Reihe Geotechnik, Heft 4, 1-225.

[45] Roscoe, K.H., Schofield, A.N., and Wroth, E.P. (1958). On the yielding of soils. Géotechnique, 8: 22-53.

[46] Saje, M. (1974). Discussion. Proc. 4th Danube European Conf. Soil Mech. Found. Eng., Bled, II: 82-83.

[47] Sandhu, R.S., and Wilson, E.L. (1969). Finite element analysis of seepage in elastic media. Jour. Eng. Mechs. Div., ASCE, 95, SM 1, 285-312.

[48] Schiffman, R.L., Chen, A.T.F., and Jordan, J.C. (1969). An analysis of consolidation theories. Jour. Soil. Mech. and Found. Div., ASCE, 95, SM 1, 285-312.

[49] Singh, A., and Mitchell, J.K. (1968). General stress-strain-time functions for soils. J. of the Soil Mech. Found. Div., ASCE, Vol. 94, SM 1, 21-46.

[50] Stroganov, A.S. (1963). One-dimensional deformation of soil as nonlinear visco-elastic medium. Proc. European Conf. Soil Mech. Found. Eng., Wiesbaden, I: 55-60.

[51] Šuklje, L. (1957). The analysis of the consolidation process by the isotache method. Proc. 4th Int. Conf. Soil Mech. Found. Eng., London, I: 200-206, III: 107-109.

[52] Šuklje, L. (1967). Common factors controlling the consolidation and the failure of soils. Proc. Geotechn. Conf. Oslo, I: 153-158.

[53] Šuklje, L., and Kogovšek, B. (1968). Isochrones of a uniformly loaded layer of viscous soils. III Sesja Naukowa Wydzialu Budownictwa Ladowego Politechniki Wroclawskiej 1968 r., Referaty I: 369-380.

[54] Šuklje, L. (1969-a). Rheological aspects of soil mechanics. Wiley Interscience, London, 571 p.

[55] Šuklje, L. (1969-b). Consolidation of viscous soils subjected to continuously increasing uniform load. "New Advances in Soil Mechanics", Czechoslovak Scientific and Technical Society, Prague, I: 199-235.

[56] Šuklje, L., and Simončič, M. (1972). The use of isotaches in the numerical analysis of radial consolidation. University of Ljubljana, Acta Geotechnica, No. 41: 1-57.

[57] Šuklje, L. (1972, 1973). Discussions to J.E. Garlanger's paper: The consolidation of soils exhibiting creep under constant effective stress. Géotechnique 22: 670-673, Géotechnique 23: 283-284.

[58] Šuklje, L. (1973). The use of isotaches in the consolidation analysis. Proc. 8th Int. Conf. Soil Mech. Found. Eng., Moscow, Vol. 4.3: 116, 62-63.

[59] Šuklje, L., and Kozak, J. (1974). Consolidation of partly saturated viscous soils. University of Ljubljana, Acta Geotechnica, No. 54: 1-20.

[60] Šuklje, L., and Kovačič, I. (1974). The role of the effective stress speed in the consolidation analysis. Proceedings of the 4th Danube-European Conference on Soil Mechanics and Foundation Engineering, Bled 1974, Vol. 1: 233-240.

[61] Šuklje, L., and Majes, B. (1975). Development of displacements in the viscous plane-strain space. University of Ljubljana, Acta Geotechnica, No. 58: 1-13.

[62] Šuklje, L. (1977). Stresses and strains in non-linear viscous soils. Forwarded for publication in Numerical and Analytical Method in Geomechanics, London, New York.

[63] Šuklje L., and Kovačič, I. (1978). Consolidation of drained stratified viscous soils. Forwarded for publication in the Proceedings of the Engineering Foundation Conference on Evaluation and Prediction of Subsidence, Pensacola Beach, Florida, January 1978.

[64] Tan, Tjong-Kie (1954). Onderzoekingen over de rheologische eigenschappen van klei (Investigations on the rheological properties of clay, in Dutch with English summary). Uitgeverij Excelsior, 's-Gravenhage, 152 p.

[65] Tan, Tjong-Kie (1957). Three-dimensional theory on the consolidation and flow of the clay layers. Scientia Sinica, 6, No. 1: 203-215.

[66] Taylor, D.W., and Merchant, W. (1940). A theory of clay consolidation accounting for secondary compression. J. Maths and Physics, 19: 167-185.

[67] Taylor, D.W. (1942). Research on consolidation of clays. Dept. of Civil and Sanitary Eng., MIT, Publ. Serial 82: 1 -147.

[68] Terzaghi, K. (1923). Die Berechnung der Durchlässigkeitsziffer des Tones aus dem Verlauf der hydrodynamischen Spannungserscheinungen. Sitzungsber. Akad. Wiss. Wien, mathem. -naturw. Kl., Abt. II a, 132 Bd., 3.u. 4.H.

[69] Vidmar, S. (1974). Discussion. Proc. 4th Danube-European Conf. Soil Mech. Found. Eng., Vol. II: 67-68.

[70] Vyalov, S.S. (1963). Reologiya merzlykh gruntov (Rheology of ,zen soils, in Russian), Prochnost i polzuchest merzylch gruntov. Izd. Akad. nauk SSSR, Moskva, 5-54.

[71] Zaretsky, Yu. K. (1967). Teoriya konsolidatsii gruntov (Theory of consolidation of soils, in Russian). Izd. Nauka, Moskva, 270 p.

[72] Zienkiewicz, O.C., and Naylor, D.J. (1971). The adaption of critical state soil mechanics theory for use in finite elements. Stress-strain Behaviour of Soils (The Roscoe Memorial Symposium), Cambridge University, 537-547.

Printed in the United States
by Bookmasters

Printed in the United States
By Bookmasters